**Materials for the
Engineering Technician**

by the same author

Engineering Metallurgy
Part 1: Applied Physical Metallurgy
Part 2: Metallurgical Process Technology

Properties of Engineering Materials
also of interest

Manufacturing Technology
M. Haslehurst, DIC, CENG, MIMECHE, MIPRODE

Materials for the Engineering Technician

Raymond A. Higgins BSc (Birm), C. Eng., FIM

*Senior Lecturer in Materials Science, The College of
Commerce and Technology, West Bromwich; former Chief
Metallurgist, Messrs Aston Chain and Hook Co., Ltd,
Birmingham; and Examiner in Metallurgy to The Institution
of Production Engineers, The City and Guilds of London Institute,
The Union of Lancashire and Cheshire Institutes, and The Union
of Educational Institutes*

HODDER AND STOUGHTON
LONDON SYDNEY AUCKLAND TORONTO

ISBN 0 340 15272 9

First published 1972
Reprinted 1975, 1976, 1977, 1978, 1979, 1980, 1981, 1982, 1983

Printed in Hong Kong for
Hodder and Stoughton Educational,
a division of Hodder and Stoughton Ltd.,
Mill Road, Dunton Green, Sevenoaks, Kent,
by Colorcraft Ltd

Editor's Foreword

The study of Engineering Materials (in some form) occurs in virtually every engineering technician course. In many of these courses, the topic is deliberately linked with another technology, such as Engineering Drawing. In other courses, the physical properties of materials are linked with Engineering Science; whilst the usage of those materials is associated with Design. In addition, the studies often progress through a course, commencing normally with the study of the more common steels and non-ferrous alloys in the early stages, developing into quite complex alloys and non-ferrous materials, particularly plastics materials, in the later stages.

When individual text-books are written for particular stages and/or subjects of technician courses, it is almost inevitable that a certain lack of continuity will occur. One method of overcoming this difficulty is to provide a textbook on materials which can be used as supplementary reading for any stage of any technician course. In this manner, a reader can be encouraged to occasionally view the topic as a single entity, rather than as a series of sometimes disconnected items. In addition, whilst textbooks provide a valuable service for examination preparation, they can serve an equally useful service in the field of liberalising a reader's knowledge of a technology.

The author needs little introduction, and enjoys an enviable reputation from his previous works. The style of writing and the content of this volume reflects the author's wide experience in industry, his awareness of recent advances in the technology, and his appreciation of the needs of students of engineering in general.

Whilst this book is particularly applicable to the Technician Education Council (TEC) course, it will also serve as a valuable reference work, not only for any technician course, but also as a liberalising effect on courses of the diploma and degree type, where a subject can so easily be treated in far too narrow a vein.

M. G. PAGE

Preface

My grandfather used to take trout from the River Tame, which runs alongside our new college here in West Bromwich. During my boyhood, fish no longer lived there, though occasionally an adventurous water-vole would make an exploratory dive. Today the Tame is a filthy open sewer, which will support neither animal nor vegetable life, and even the lapwings have forsaken its surrounding water-meadows. The clear-water tributaries of the Tame, where as a boy I gathered watercress for my grandmother—and tarried to fish for sticklebacks—are now submerged by the acres of concrete which constitute the link between the M5 and M6 motorways. On its way to the sea, the Tame spews industrial poison of the Black Country into the unsuspecting Trent, doing little to maintain the ecological stability of that river.

We Midlanders are not alone in achieving this kind of environmental despoilation. It seems that the Americans have succeeded in poisoning considerable areas of their Great Lakes. At this rate of 'progress', we must seriously consider the ultimate pollution of the sea, which, far from being 'cruel', provides Man with a great deal of his food. More important still, much of his supply of oxygen is generated by the vast forests of marine kelp which is generally referred to, somewhat unkindly, as 'seaweed'.

We would do well to consider the extent to which a more effective use of materials might alleviate the pollution of our environment. Suppose, for example, a motor car were constructed of materials such that it would last much longer. Presumably, since fewer motor cars would then be produced, the total amount of environmental pollution associated with their production would also be reduced. Further, since their scrap value would be higher, less of them would be left around to disfigure—or pollute—our countryside. Obviously, if such a course were followed, the whole philosophy of consumer production would need to be rethought, since a reduced total production would mean underemployment per worker (judged by present standards, rather than those which will obtain during the twenty-first century). However, these are problems to be solved by the planner and the politician, rather than by the mechanical engineer.

Pollution of the environment is, however, only one factor in the use of materials that the engineer must consider. The supplies of basic raw materials are by no means inexhaustible, and, by the end of the present century, reserves of metals like copper and tungsten will have run out, if we continue to gobble them up at present rates. In the more distant future it may well become a punishable offence to allow iron in any of its forms to rust away and become so dispersed amongst its surroundings that it cannot conveniently be reclaimed. The motor car of the future will need to be made so that it does not corrode. Such an automobile, made to last say forty years, would inevitably cost a lot of money, but, when worn out, would be returned to the factory for 'dismemberment', so that almost 100 per cent of its basic materials would be reclaimed. Only in this way will looting and pollution of the environment be reduced to tolerable proportions, particularly if the world population continues to increase even at only a fraction of its present rate.

It follows that a greater appreciation of the properties of all available materials is necessary to the mechanical engineer, and, although this book has been written with the mechanical engineering technician specifically in mind, it is hoped that it will provide a useful introduction to Materials Science for others engaged in engineering production. The book has been so written that only a most elementary knowledge of general science is necessary in order that it can be read with advantage. As a point of interest, the original idea of the work was conceived when the author was blizzard-bound in his caravan one Easter mountaineering holiday in Skye. It was finished in high summer in the Haute Savoie, when plagues of vicious insects in the alpine meadows made it equally unpleasant to venture forth.

*　　*　　*

Since writing the above paragraphs, I have learned that plans to clean up our River Tame are now 'under consideration', so we may get some action —eventually. In the meantime, the luckless river continues to pursue its fetid course outside the window of the lecture room where I teach. At 9.0 a.m. on Monday morning, the water is clear enough for me to discern an old motor tyre, presumably thrown there by small boys. By midday, the water has assumed a particularly disagreeable appearance, and the tyre has disappeared in a 'soup' of nacreous hue. As the working week proceeds, the river changes in colour from time to time, as the products of our 'effluent society' are discharged into it. Overhead, a hopeful kestrel still hunts this poisoned land; I wish him luck.

R. A. Higgins
The College of Commerce and Technology
West Bromwich, Staffs

Contents

Chapter One
Setting the Stage

1.10 During the period in which William Shakespeare was writing his Sonnets, only half a dozen metals were known to Man. Nevertheless, Man had come a long way since *Pithecanthropus erectus* finally raised himself from a four-legged posture, and stood firmly and permanently upon his hind-legs. True, this had taken quite a long time—probably half a million years or so—and on this sort of time-scale Ancient Greece seems but yesterday, and the beginning of the Renaissance a few hours since.

The Renaissance was, above all, a period when fine arts flourished, and the spirit of Man began to rise again from the Dark Ages which for a brief thousand years had engulfed him. It was the day of Michelangelo, Cellini, Erasmus, and Leonardo da Vinci, but, though Leonardo is remembered specifically as a painter, he was also something of a scientist and an engineer. In fact it is reported that he once designed a flying-machine which, given suitable motive power, might well have flown. Had he lived today, Leonardo would undoubtedly have been a designer of space-ships, rather than devoting his great energy to art. As to whether mankind would be better or worse off for this is perhaps another question: certainly we would be without either his great masterpiece the enigmatic 'Mona Lisa', or his portrayal of 'The Last Supper'.

1.11 This great upsurge of human activity took place during the Renaissance, even though Man had surprisingly few materials with which to work. Apart from the half-dozen metals mentioned above, he had stone and clay, wood and straw, and little else from which to construct anything. As our immortal bard wrote his Sonnets, living in a house built almost entirely of wood, bricks, thatch—and mud, little did he know that, in the winking of an eye, Merrie England would vanish, probably for ever, and that the Industrial Revolution and the age of technology were about to begin.

1.20 We are now living in the second half of the twentieth century, and the purpose of this book is to describe, in very simple terms, the nature, properties, and uses of the more important engineering materials in use today.

The scientist has long since discovered all of the metals which can exist in the Earth's crust—or elsewhere for that matter—and we now have a total of seventy of them (not counting the man-made products of nuclear physics). True, of these seventy metallic elements, some are extremely rare, and only about half are of any possible use to the engineer. Caesium, for example, is a very light metal, but is extremely soft and—worst still—

oxidises spontaneously, with a flash of light, as soon as it comes into contact with the atmosphere. However, knowledge of these apparently useless metals often helped the chemist of last century in his search for 'missing' elements, the existence of which was then only suspected. Moreover, though chemical elements in some groups may at present have no commercial value, the same was true of uranium before 1940.

Knowledge of non-metallic materials, too, has developed with increasing impetus during the last century. Plastics and reinforced concrete play an increasing part in our daily lives; whilst it seems likely that the element carbon will soon form the basis of our strongest materials.

1.21 Metals can be roughly divided into two groups—the ferrous metals and the non-ferrous metals. The former are materials based on the metal iron, and include wrought iron, cast iron, and the various types of steel. The non-ferrous metals comprise all the rest—aluminium, copper, lead, zinc, tin, and some thirty others which are of value in engineering.

1.22 When considering the non-metallic materials which are of interest to the engineer, we must not forget that, although man-made materials like concrete and plastics are very important, great use is still made of the naturally occurring substances like clay, wood, diamond, graphite, rubber, asbestos, and crude oils.

These, then, are the 'actors in our play'—the *dramatis personae*, so to speak.

Chapter Two
The Crystal Structure of Metals

2.10 Atoms are very tiny particles indeed. A pin-head contains about 350 000 000 000 000 000 of them—give or take a few millions—yet we now know that these small particles are themselves built of even smaller units, the most important of which are negatively-charged electrons and positively-charged protons. All solid materials consist of atoms arranged in some pattern peculiar to that material. The atoms are held together by electrical forces, basically of attraction between the negatively-charged electrons of one atom and the positively-charged protons of its neighbours.

Altogether there are ninety-two different types of atom which occur naturally. The smallest and simplest is that of hydrogen; whilst the largest, some two-hundred and thirty-eight times as massive, is that of uranium. A chemical *element* contains atoms all of one type. Thus there are ninety-two different chemical elements, of which over seventy are metals. Some of these metallic elements are extremely rare; whilst others are virtually useless to the engineer. Consequently, less than twenty of them (table 2.1) are in common use in engineering alloys.

2.11 All metals—and other elements, for that matter—can exist as either gases, liquids, or solids. The 'state' in which a metal exists depends upon the conditions of temperature and pressure which prevail at the time. Thus mercury will freeze to form a solid, rather like lead, if cooled to −39°C; and will boil to form a gas or vapour if heated to 357°C at atmospheric pressure. (Metals with low boiling-points—such as mercury, cadmium, and zinc—can be purified by distillation.) At the other end of the scale, tungsten melts at 3410°C, and vaporises at 5930°C.

2.12 In a liquid metal, the atoms are jumbled together willy-nilly, and, since they are held together only by weak forces of attraction at this stage, the liquid lacks cohesion, and will flow. When the metal solidifies, the energy of each atom is reduced. This energy is given out as latent heat during the solidification process, which for a pure metal occurs at a fixed temperature (fig. 2.1). During solidification, the atoms arrange themselves according to some regular pattern, or 'lattice structure'. Each atom becomes firmly bonded to its neighbours by stronger forces of attraction; so the solid metal acquires strength.

Since the atoms are now arranged in a regular pattern, they generally occupy less space. Thus most metals shrink during solidification, and the foundryman must allow for this, not only by making the wooden pattern a little larger than the required size of the casting, but also by providing adequate runners and risers, so that molten metal can feed into the body

Metal	Chemical symbol	Melting-point (°C)	Properties and uses
Aluminium	Al	659·7	The most widely used of the 'light metals'. Common in the Earth's crust.
Antimony	Sb	630·5	A very brittle metal, which is used mainly in bearing alloys and type-metal.
Beryllium	Be	1285	A light metal used to harden copper, and for nuclear-power equipment. Would be widely used in aircraft and spacecraft were it not so scarce and hence very expensive.
Cadmium	Cd	320·9	Sometimes used to plate other metals, and to strengthen copper telephone-wires.
Chromium	Cr	1890	Its high resistance to corrosion makes it useful for plating other metals, and as an addition to stainless steels.
Cobalt	Co	1495	Used in permanent magnets and high-speed steels.
Copper	Cu	1083	Used mainly in the electrical industries, because of its high conductivity. Also in bronzes, brasses, and cupro-nickel.
Iron	Fe	1535	A fairly soft metal when pure, but hard and strong when alloyed to form steel.
Lead	Pb	327·4	A soft, heavy metal, but not the densest, as the phrase 'as heavy as lead' suggests.
Magnesium	Mg	651	Used along with aluminium in the 'light alloys'.
Manganese	Mn	1260	Added to most steels as a deoxidant before the steel is cast.
Molybdenum	Mo	2620	A heavy metal used in high-speed and other alloy steels.
Nickel	Ni	1458	Used to toughen steels and many non-ferrous alloys; also for plating.
Niobium	Nb	1950	Small amounts used in some steels and aluminium alloys; also in nuclear-power equipment. Formerly known as 'columbium' in the USA.
Tin	Sn	231·9	A rather expensive metal. 'Tin cans' carry only a very thin coating of tin on mild steel. Very resistant to corrosion.
Tungsten	W	3410	Its very high melting-point makes it useful for electric lamp filaments. Also used in high-speed and heat-resisting steels.
Vanadium	V	1710	Used as a hardener in some alloy steels; also in nuclear-power equipment.
Zinc	Zn	419·5	Used widely for galvanising mild steel, and as a basis for a group of die-casting alloys ('Mazak'). Brasses are copper-zinc alloys.

Table 2.1—Metals commonly used in engineering.

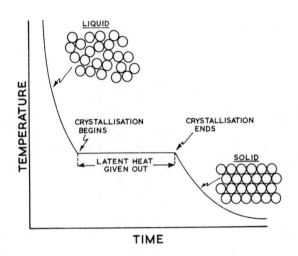

Fig. 2.1—A pure metal solidifies at a fixed single temperature, and the atoms arrange themselves in some regular pattern.

of a casting as it solidifies, and so prevent the formation of internal shrinkage cavities.

2.13 There are three different patterns in which the atoms of the more important metals arrange themselves on crystallisation (fig. 2.2A). It should be appreciated that the upper diagrams represent the simplest unit in each case. Here the positions occupied by the centres of atoms are shown. In fact each outer 'face' of the figure is also part of the next adjacent unit, as shown in the lower diagrams (fig. 2.2B). These lattice structures were derived by means of X-ray analysis.

2.14 Of these 'space lattices', the close-packed hexagonal arrangement represents the closest packing of atoms. It is the arrangement which would be produced if a second layer of snooker balls were allowed to fall into position on top of a set already packed in the triangle. The face-centred cubic arrangement is also a fairly close packing of atoms, but the body-centred cubic form is relatively 'open'; that is, the lattice contains relatively more free space.

Iron is said to be an *allotropic* element, because it can exist in more than one crystalline form. The body-centred cubic form (α-iron) exists up to 910°C, when it changes to the face-centred cubic form (γ-iron). On cooling again, the structure reverts to body-centred cubic. It is this fact which enables steel to be heat-treated in its own special way, as we shall see later in this book (12.21). Unfortunately, the sudden volume change which occurs as γ-iron changes to α-iron on being quenched gives rise to the

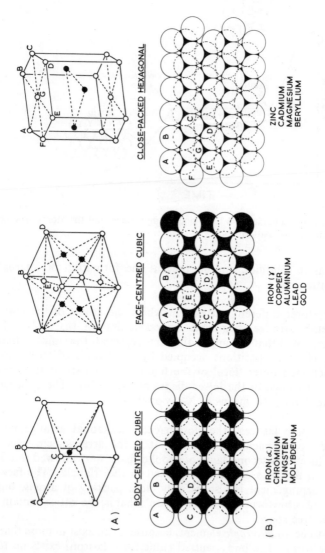

Fig. 2.2—The principal types of lattice structure occurring in metals.
(A) indicates the positions of the centres of atoms only, in the simplest unit of the structure; whilst (B) shows how these units occur in a continuous crystal structure viewed in 'plan'. (In each case the letters A, B, C, etc. indicate the appropriate atoms in both diagrams.) Although the atoms here are shown black or white, to indicate in which layer they are situated, they are, of course, all of the same type.

formation of internal stresses, and sometimes distortion or even cracking of the component. However, we must not look a gift-horse in the mouth: but for this freak of nature in the form of the allotropic change in iron, we would not be able to harden steel. Without steel, as Man's most necessary engineering material, we would still, presumably, be in the Bronze Age.

Dendritic solidification

2.20 When the temperature of a molten pure metal falls to its freezing-point, crystallisation will begin. The nucleus of each crystal will be a single unit of the appropriate crystal lattice. For example, in the case of a metal with a body-centred cubic lattice, nine atoms will come together to form a single unit, and this will grow as further atoms join the lattice structure (fig. 2.3). These atoms will join the 'seed crystal' so that it grows most

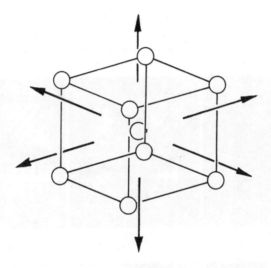

Fig. 2.3—The nucleus of a metallic crystal (in this case a body-centred cubic structure).

quickly in those directions in which heat is flowing away most rapidly. Soon the tiny crystal will reach visible size, and form what is called a 'dendrite' (fig. 2.4). Secondary and tertiary arms develop from the main 'backbone' of the dendrite—rather like the branches and twigs which develop from the trunk of a tree, except that the branches in a dendrite follow a regular geometrical pattern. The term 'dendrite' is, in fact, derived from the Greek *dendron*—a tree.

2.21 The arms of the dendrite continue to grow until they make contact with the outer arms of other dendrites growing in a similar manner near by. When the outward growth is thus restricted, the existing arms thicken

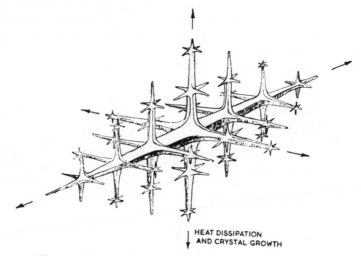

Fig. 2.4—The early stages in the growth of a metallic dendrite.

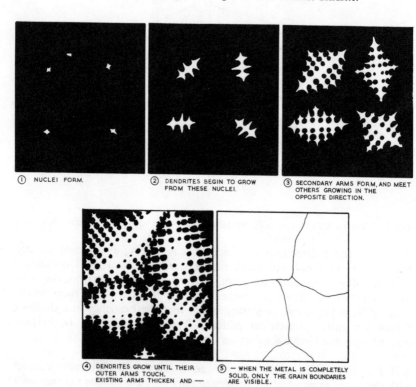

① NUCLEI FORM.

② DENDRITES BEGIN TO GROW FROM THESE NUCLEI.

③ SECONDARY ARMS FORM, AND MEET OTHERS GROWING IN THE OPPOSITE DIRECTION.

④ DENDRITES GROW UNTIL THEIR OUTER ARMS TOUCH. EXISTING ARMS THICKEN AND —

⑤ — WHEN THE METAL IS COMPLETELY SOLID, ONLY THE GRAIN BOUNDARIES ARE VISIBLE.

Fig. 2.5—The dendritic solidification of a metal.

until the spaces between them are filled, or, alternatively, until all the remaining liquid is used up. As mentioned earlier in this chapter, shrinkage usually accompanies solidification, and so liquid metal will be drawn in from elsewhere to fill the space formed as a dendrite grows. If this is not possible, then small shrinkage cavities are likely to form between the dendrite arms.

2.22 Since each dendrite forms independently, it follows that outer arms of neighbouring dendrites are likely to make contact with each other at irregular angles, and this leads to the irregular overall shape of crystals, as indicated in fig. 2.5. In a similar manner, the trees of a forest push and jostle each other as they reach towards the light, so that forest trees are rarely of pleasing, regular shapes.

2.23 If the metal we have been considering is pure, then there will be no hint of the dendritic process of crystallisation once solidification is complete, because all atoms in a pure metal are identical. If impurities were dissolved in the molten metal, however, these would tend to remain in solution until solidification was almost complete. They would therefore remain concentrated in that metal which solidified last; that is, between the dendrite arms. In this manner they would reveal the dendritic pattern (fig. 2.6) when a suitably prepared specimen was viewed under the microscope. This concentration or *segregation* of impurities at crystal boundaries

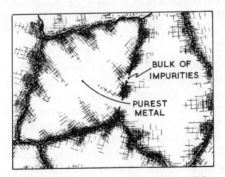

Fig. 2.6—The local segregation of impurities.
The heavily shaded region near the crystal boundaries contains the bulk of the impurities.

explains why a small amount of impurity can have such a devastating effect on mechanical properties, making the cast metal brittle and likely to fail along the crystal boundaries. In addition to this local segregation at all crystal boundaries, there is a general accumulation of impurities in the central 'pipe' of a cast ingot (fig. 2.7). This is where metal solidifies last of all, and has become most charged with impurities; relatively pure metal having crystallised during the early stages of solidification.

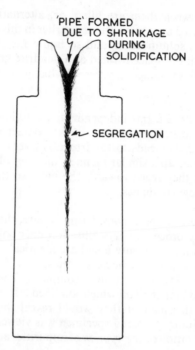

'PIPE' FORMED
DUE TO SHRINKAGE
DURING
SOLIDIFICATION

SEGREGATION

Fig. 2.7—The segregation of impurities in the central 'pipe' of an ingot.

In casting steel ingots, the central pipe is confined by using a fireclay collar on top of the mould (fig. 3.1). The mould is tapered upwards, so that it can be lifted off the solid ingot.

2.24 The rate at which a molten metal solidifies affects the size of the crystals which form. A gradual fall in temperature results in the formation of only a few nuclei, and so the crystals can grow unimpeded to a large size. Rapid cooling, however, leads to the formation of a sudden 'shower' of nuclei. Since the resultant crystals are large in number, they will be small in size. As the foundryman says, 'Chilling causes fine-*grain** castings'. Because of the difference in the rates of cooling, the resultant grain size of a die-casting is small as compared with that of a sand-casting. This is an advantage, since fine-grained castings are generally tougher and stronger than those with a coarse grain size.

2.25 When a large ingot solidifies, the rate of cooling varies from the outer skin to the core during the crystallisation process. At the onset of crystallisation, the cold ingot-mould chills the molten metal adjacent to it; so a layer of small crystals is formed. Due to the heat flow outwards, the mould warms up, and so chilling becomes less severe. This favours the growth inwards of elongated or columnar crystals. These grow inwards more

* The term 'grain' is often used to mean 'crystal'.

quickly than fresh nuclei can form, and this results in their elongated shape. The residue of molten metal, at the centre, cools so slowly that very few nuclei form, and so the crystals in that region are relatively large. They are termed 'equi-axed crystals'—literally 'of equal axes' (fig. 2.8).

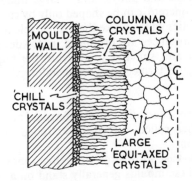

Fig. 2.8—Zones of different crystal forms in an ingot.

2.26 A cast material tends to be rather brittle, because of both its coarse grain and the segregation of impurities mentioned above. When cast ingots are rolled or forged to some other shape, the mechanical properties of the material are vastly improved during the process, for reasons discussed later (6.50). Frequently, the engineer uses sand- or die-castings as integral parts of a machine. Often no other method is available for producing an intricate shape, though in many cases a sand-casting will be used because it involves the *cheapest* method of producing a given shape. The choice rests with the production engineer: he must argue the mechanical and physical properties required in a component with the Spectre of Production Costs—who is for ever looking over his shoulder.

Chapter Three
Casting Processes

3.10 Most metallic materials are melted and then cast—either into some final shape or, alternatively, into ingot form for subsequent mechanical working. Some metals and alloys, and a large number of non-metallic substances such as concrete, can be shaped only by casting, since they lack the necessary properties which make them amenable to working processes.

A few metallic substances are produced in powder form. The powder is then compressed and sintered to provide the required shape. This branch of technology is termed 'powder-metallurgy' (7.40).

Ingot-casting

3.20 Many alloys, both ferrous and non-ferrous, are cast in the form of ingots which are then rolled, forged, or extruded into strip, sheet, rod, tube, or other sections.

Steel is generally cast into large, iron ingot-moulds holding several tonnes of metal. These moulds generally stand on a flat metal base, and are tapered upwards very slightly so that the mould can be lifted clear of the solid ingot. A 'hot top' (fig. 3.1) is often used in order that a reservoir of molten metal shall be preserved until solidification of the body of the ingot is complete. This reservoir feeds metal into the 'pipe' which forms as the main body of the ingot solidifies and contracts. Many non-ferrous metals are cast as slabs in cast-iron 'book-form' moulds; whilst some are cast as cylindrical ingots for subsequent extrusion.

3.21 *Continuous casting.* This process is used to produce ingots of both ferrous and non-ferrous alloys (fig. 3.2). Here the molten metal is cast into a short water-cooled mould, A, which has a retractable base, B. As solidification begins, the base is withdrawn downwards at a rate which will keep pace with that of pouring.

With ordinary ingots, a portion of the top must always be cropped off and rejected, since it contains the pipe, and is a region rich in impurities. In continuous casting, however, there is little process scrap, since very long ingots are produced, and consequently there is proportionally less rejected pipe.

Sand-casting

3.30 The production of a desired shape by a sand-casting process first involves moulding foundry sand around a suitable pattern, in such a way that the pattern can be withdrawn to leave a cavity of the correct shape in the sand. To facilitate this, the sand mould is split into two or more parts, which can be separated so that the wooden pattern may be removed.

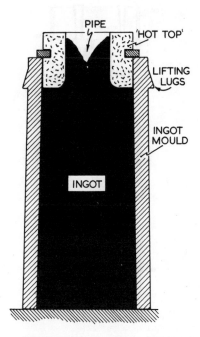

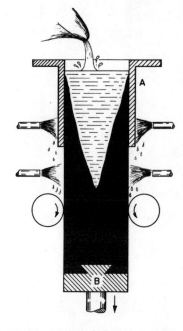

Fig. 3.1—A typical mould for producing steel ingots. The 'hot top' restricts the formation of the 'pipe' which is caused by contraction of the metal as it solidifies.

Fig. 3.2—A method for continuous casting.

3.31 The production of a very simple sand mould is shown in fig. 3.3. The pattern of the simple gear blank is first laid on a moulding-board, along with the 'drag' half of a moulding-box (i). Moulding sand is now riddled over the pattern, and rammed sufficiently for its particles to adhere to each other. When the drag has been filled, the sand is 'cut' level with the edge of the box (ii), and the assembly is turned over (iii).

A layer of parting sand (dry, clay-free material) is now sifted on to the sand surface, so that the upper half of the mould will not adhere to it when this is subsequently made. The 'cope' half of the moulding-box is now placed in position, along with the 'runner' and 'riser' pins, which are held steady by means of a small amount of moulding sand pressed around them (iv). The purpose of the runner in the finished mould is to admit the molten metal; whilst the riser provides a reservoir from which molten metal can feed back into the casting as it solidifies and shrinks.

Moulding sand is now riddled into the cope, and rammed around the pattern, runner, and riser (v). The cope is then gently lifted off, the pattern is removed, and the cope is replaced in position (vi), so that the finished mould is ready to receive its charge of molten metal.

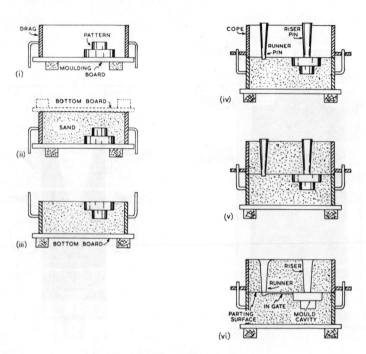

Fig. 3.3—Moulding with a simple pattern.

The foregoing description deals with the manufacture of a sand mould of the simplest type. In practice, a pattern may be of such a complex shape that the mould must be split into several sections, and consequently a multi-part box is used. *Cores* may be required to form holes or cavities in the casting.

3.32 Sand-casting is a very useful process, since very intricate shapes can be produced in a large range of metals and alloys. Moreover, relatively small numbers of castings can be made economically, since the necessary outlay on the simple equipment required is low. Wooden patterns are also cheap to produce, as compared with the metal die which is necessary in die-casting processes.

Die-casting
3.40 In die-casting, a permanent metal mould is used, and the charge of molten metal is either allowed to run in under the action of gravity (gravity die-casting), or is forced in under pressure (pressure die-casting). A number of different types of machine is employed in pressure die-casting, but possibly the most widely used is the cold-chamber machine (fig. 3.4). Here a charge of molten metal is forced into the die by means of a plunger. As soon as the casting is solid, the moving platen is retracted, and as it comes

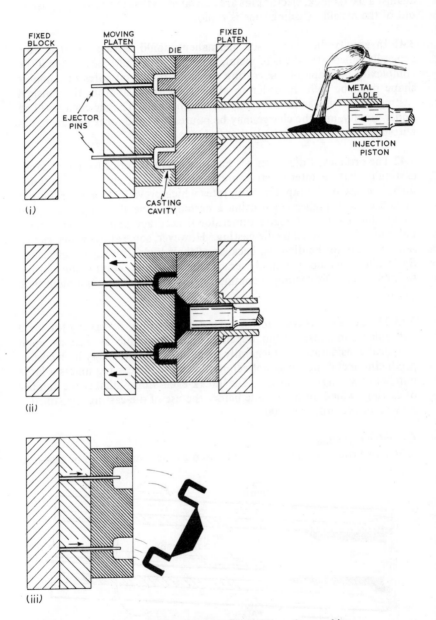

Fig. 3.4—A 'cold-chamber' pressure die-casting machine.
The molten metal is forced into the die by means of the piston. When the casting is
solid, the die is opened, and the casting moves away with the moving platen, from which
it is ejected by 'pins' passing through the platen.

against a fixed block, ejector pins are activated so that the casting is pushed out of the mould. Cycling time is rapid.

3.41 In gravity die-casting—or 'permanent-mould casting', as it is now generally called—the die is of metal, and may be of multi-part design if the complexity of shape of the casting demands it. Metal cores of complex shape must be split, in order to allow their removal from the finished casting; otherwise, sand cores may have to be used. The die cavity is filled under gravity, and the charge may be poured by hand, or it may be fed in automatically in modern high-speed plant.

3.42 The product of die-casting is metallurgically superior to that of sand-casting in that the internal structure is more uniform, and the grain much finer, because of the rapid cooling rates which prevail. Moreover, output rates are much higher when using a permanent metal mould than when using sand moulds. Greater dimensional accuracy and a better surface finish are also obtained by die-casting. However, some alloys which can be sand-cast cannot be die-cast, because of their high shrinkage coefficients. Such alloys would inevitably crack due to their contraction during solidification. Die-casting is confined mainly to zinc- or aluminium-base alloys.

3.43 The decision whether to use a sand-casting or a die-casting for some particular component often depends upon economic as well as upon technical considerations. Generally speaking, due to the very high cost of producing metal dies, it is not economic to use die-casting unless a large number of castings is required. The 'break-even' value—that is, the number of castings which must be made before the use of die-casting is justified—may be of the order of 5000.

Centrifugal casting
3.50 This process is most commonly used in the manufacture of cast-iron

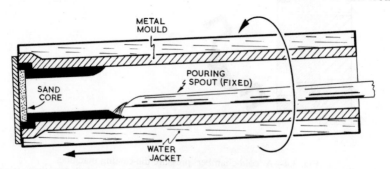

Fig. 3.5—The principle of the centrifugal casting of pipes.
The mould rotates and also moves to the left, so that molten metal is distributed along the mould surface.

pipes for water, sewage, and gas mains. A permanent cylindrical metal mould, *without* any central core, is spun at high speed, and molten metal is poured into it (fig. 3.5). Centrifugal force flings the metal to the surface of the mould; thus producing a hollow cylinder of uniform wall thickness. The product has a uniformly fine-grained outer surface, and is considered superior to a similar shape which has been sand-cast. In addition to the types of pipe mentioned above, hollow cylinders from which cast-iron piston rings can be cut are also centrifugally cast.

Shell-moulding

3.60 This process was developed in Germany by Johannes Croning during the Second World War. It was originally employed in the manufacture of hand-grenades, but has subsequently been used for the production of large numbers of castings requiring great dimensional accuracy.

3.61 It is fundamentally a sand-casting process in which the clay bond present in ordinary foundry sand is replaced by an artificial bonding material of the phenol-formaldehyde or urea-formaldehyde type (23.712). Fine sands, free of clay, are mixed with about 5% by weight of the plastic bonding-agent.

3.62 Each half of the shell mould is made on a pattern plate, which must be of metal, because of the relatively high temperatures at which the bonding-agent 'sets'. The plate is first heated to about 250°C, and is then coated with silicone oil, in order to facilitate the subsequent stripping of the shell from the pattern. The pattern plate is then placed on top of a dump-box containing the sand-resin mixture (fig. 3.6 (i)), and the box is inverted, so that the pattern becomes covered with the sand-resin mixture (ii).

The resin melts, and in approximately 30 seconds the pattern has become coated with a shell of resin-bonded sand. The shell soon becomes quite hard, because the phenolic-type resins used are of the thermosetting variety (23.45). The dump-box is turned back to its original position, so that surplus moulding material falls back into the bottom of the box (iii).

The pattern plate is then removed, and, with the shell still adhering to it, it is transferred to an oven, where the shell is hardened still further by curing for about 2 minutes at 315°C. The shell is then stripped from the pattern plate by means of ejector pins built into the plate. Normally these pins bear on the surface of the shell at the edges of the mould cavity (iv), so as not to damage the surface of the mould cavity itself.

The two halves of the mould so produced are then joined together by adhesives, bolts, clamps, or—in the case of small moulds—by spring paper-clips. Sometimes the mould is supported with metal shot, coarse sand, or gravel (v), and is then ready to receive the molten charge.

The main advantage in using shell-moulding lies in the high dimensional accuracy of the product. Tolerances of the order of 5 mm/metre are claimed to be common practice, as compared with about 15 mm/metre in

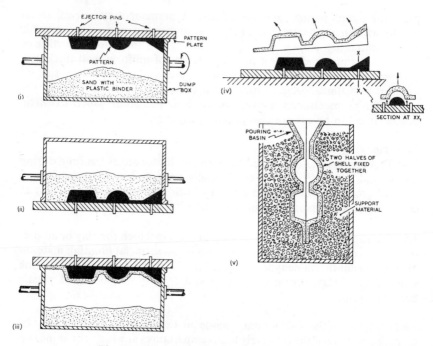

Fig. 3.6—Stages in the production of a shell-mould.

the average pressure die-casting. Moreover, the surface finish is far superior to that on an ordinary sand-casting, since loose sand and sand inclusions are absent. Since a shell mould has a small heat capacity, as compared with a sand mould, the metal can be cast at a lower temperature. For this reason, oxidation losses and gas absorption during melting are reduced to a minimum. Shell moulds will store well, so mould-making plant need not be sited near to the foundry. Labour costs too are lower than with green-sand moulding, and working conditions are much cleaner than with ordinary foundry sand. Most foundry metals can be successfully cast into shell moulds.

3.63 The principal disadvantage of shell-moulding lies in the high cost of the metal patterns, mainly because these need to be of high dimensional accuracy. Obviously this disadvantage is largely offset by the consequent greater value of the resultant casting.

Investment-casting
3.70 Although investment-casting came into public eminence fairly recently, in connection with the production of turbine blades used in jet engines, it is in fact the most ancient casting process in use. It is thought that pre-historic Man had learned to fashion an image from beeswax, and then

knead a clay mould around it. The mould was then hardened by firing, a process which also melted out the wax pattern, leaving a mould cavity without cores or parting lines.

3.71 The process was rediscovered in the sixteenth century by Benvenuto Cellini, who used it to produce many works of art in gold and silver. Cellini kept the process secret; so it was lost again until, during the latter part of last century, it was rediscovered for a second time, and became known as the *cire perdue*, or 'lost wax', process. Since an expendable wax pattern is required for the production of each mould, it follows that a permanent mould must first be produced to manufacture the wax patterns—assuming of course that a large number of similar components is required. This master *mould* could be machined in steel, or produced by casting a low melting-point alloy around a master *pattern*.

3.72 To produce wax patterns, the two halves of the mould are clamped together, and the molten wax is injected at a pressure of about $3 \cdot 5$ N/mm^2. When the wax pattern has solidified, it is removed from the mould, and the wax 'gate' is suitably trimmed (fig. 3.7 (ii)), using a heated hand-tool, so that it can be attached to a central 'runner' (iii). The assembled runner, with its 'tree', of patterns is then fixed to a flat bottom-plate by a blob of molten wax. A metal flask, lined with waxed paper and open at each end, is placed over the assembly. The gap between the end of the flask and the bottom-plate is sealed with wax, and the investment material is then poured into the flask. This stage of the process is conducted on a vibrating base, so that any entrapped air bubbles or excess moisture are brought to the surface whilst solidification is taking place (iv).

For castings made at low temperatures, an investment mixture composed of very fine silica sand and plaster of Paris is still sometimes used. A more refractory investment material consists of a mixture of fine 'sillimanite' sand and ethyl silicate. During moulding and subsequent firing of the mould, ethyl silicate decomposes to form silica, which knits the existing sand particles together, to give a strong, rigid mould.

The investment is allowed to dry in air for some eight hours. The baseplate is then detached, and the inverted flask is passed through an oven at about 150°C, so that the wax melts and runs out, leaving a mould cavity in the investment material (v). When most of the wax has been removed, the mould is pre-heated prior to receiving its charge of molten metal. The pre-heating temperature varies with the metal being cast, but is usually between 700 and 1000°C. The object of pre-heating is to remove the last traces of wax by volatilisation, to complete the decomposition of the ethyl silicate bond to silica, and also to ensure that the cast metal will not be chilled, but will flow into every detail of the mould cavity.

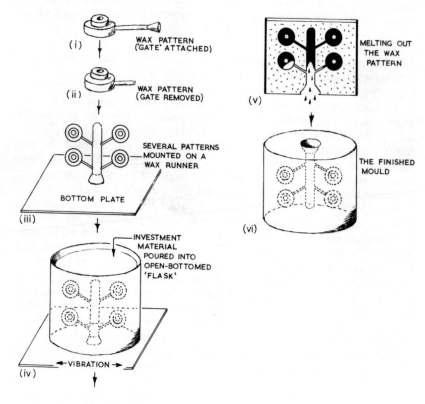

Fig. 3.7—The investment-casting (or 'lost wax') process.

3.73 Molten metal may be cast into the mould under gravity, but, if thin sections are to be formed, then it will be necessary to inject the molten metal under pressure.

The process is particularly useful in the manufacture of small components from metals and alloys which cannot be shaped by forging and machining operations. Amongst the best-known examples of this type of application are the blades of gas-turbine and jet engines.

3.74 Tolerances in the region of 3 to 5 mm/metre are obtained industrially by the investment-casting process, but it can also be used to produce complicated shapes which would be very difficult to obtain by other casting processes. A further advantage is the absence of a disfiguring parting line, which always appears on a casting made by any process involving the use of a two-part mould.

The main disadvantages of investment-casting are its high cost, and the fact that the size of components is normally limited to 2 kg or so.

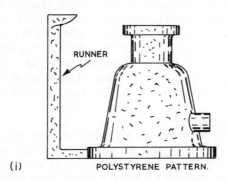

(i) POLYSTYRENE PATTERN.

(ii) PATTERN MOULDED IN SINGLE-PART FLASK.

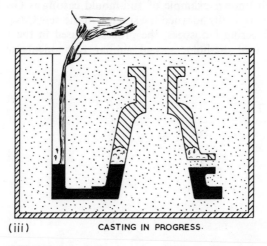

(iii) CASTING IN PROGRESS.

Fig. 3.8—The 'full-mould' process, which uses an expendable polystyrene pattern.

3.75 In addition to jet-engine blades, other typical investment-cast products include special alloy parts used in chemical engineering; valves and fittings for oil-refining plant; machine parts used in the production of modern textiles; tool and die applications, such as milling-cutters, precision gauges and forming and swaging dies; and parts for various industrial and domestic equipment, such as cams, levers, spray nozzles, food-processing plant, parts for sewing machines and washing machines.

The full-mould process

3.80 This somewhat novel process resembles investment-casting in that a single-part flask is used, so that no parting lines—and hence no fins—appear on the finished casting. However, it is essentially a 'one-off' process, since the consumable pattern is carved from expanded polystyrene (23.627). This is a polymer derived from benzene and ethylene, and in its 'expanded' form it contains only 2% actual solid polystyrene. Readers will be familiar with the substance, which is used in the manufacture of ceiling tiles, and also as a packaging material for fragile equipment. An expendable pattern, complete with runners and risers, is cut from expanded polystyrene (fig. 3.8 (i)), and is completely surrounded with sand (ii). The molten metal is then poured *on to the pattern*, which melts and burns very quickly, leaving a cavity which is immediately occupied by the molten metal. No solid residue is formed, and the carbon dioxide and water vapour evolved in the combustion of the polystyrene do not dissolve in the molten metal, but escape through the permeable mould. Moulding can be achieved merely by pouring sea sand around the pattern. As the polystyrene burns, it produces a tacky bond between the sand grains for long enough for a skin of metal to form.

3.81 A well known example of full-mould casting is Geoffrey Clarke's cross which originally adorned the summit of the new Coventry Cathedral. In the engineering industries, the process is used in the manufacture of large press-tool die-holders, and similar components in the 'one-off' category.

Chapter Four
Mechanical Testing

4.10 If asked what he thought was the most important property of steel, the average layman would no doubt refer to its 'strength'. To a metallurgist, however, the most significant mechanical properties of a metal relate to the amount it can be deformed before it fractures, and in this instance the method of causing such deformation must be considered.

4.11 We say that copper is a very *malleable* metal, meaning that it can be deformed a great deal by *compression* before it shows signs of cracking. Malleable metals can be rolled, forged, or extruded, since these are all processes where the metal is shaped under pressure. Malleability generally increases with temperature, and so processes involving pressure are invariably hot-working processes; that is, they are carried out on heated ingots or slabs of metal.

Copper is also very *ductile*; that is, it can be deformed considerably by *tension* before it fractures. Whilst all ductile metals are malleable, the converse is not always true. Some metals, although soft, are also weak in tension, and therefore tend to tear apart whilst being stretched. The ductility of all metals decreases as the temperature rises, because they are weaker at high temperatures. Both malleability and ductility are reduced by the presence of impurities in the metal or alloy.

4.12 The *toughness* of a metal refers to its ability to withstand bending without fracture. Copper is a very tough material, since a piece of copper wire can be bent to and fro many times before it will fracture. Thus copper is very tough; whilst cast iron possesses the opposite property—brittleness.

These properties of malleability, ductility, and toughness are fundamental properties which are not easy to measure in simple numerical terms; all we can do is to arrange the metals in order of the property concerned. This is of very little use to the engineer. He is primarily concerned with the balancing of *forces* when he designs some structure or machine; he must therefore know the effects which the application of forces may have on a material, before he can use it.

4.13 Various mechanical tests have therefore been devised over the years, with the object of comparing the amount of deformation produced in a metal with the force which was employed to produce it. Thus, in a tensile test we measure the force required to stretch a specimen of the metal until it breaks; whilst in various hardness tests we produce a small dent in the surface of a test-piece by using a compressive force. The hardness number is then calculated as the force used divided by the surface area of the impression produced by it.

4.14 Complete information on the behaviour of metals and alloys when subjected to these derived tests has been compiled by such bodies as the British Standards Institution, who have drawn up their series of BS specifications. It is on information such as this that the engineer bases his designs, and accepts his materials.

4.15 In the type of test mentioned above, the test-piece is destroyed during the testing process. Such tests are therefore known as *destructive tests*, and can only be applied to individual test-pieces. These are taken from a batch of material which is proposed for use for some specific purpose, and they are therefore assumed to be representative of the batch. Tests of a different nature and purpose are used to examine manufactured components for internal flaws and faults; for example X-rays are used to seek internal cavities in castings. These tests are generally referred to as *non-destructive tests* (N.D.T.), since the component, so to speak, 'lives to tell the tale'.

The tensile test

4.20 The tensile strength of a material is the stress required to cause fracture of a test-piece in tension. Many readers will be familiar with tensile-testing methods; nevertheless, a brief outline of these methods will be given here.

A test-piece of known cross-sectional area is gripped in the jaws of a testing-machine, and is subjected to a tensile force which is increased by suitable increments. For each increment of force, the amount by which the length of a known 'gauge length' on the test-piece increases is measured using a suitable extensometer. When the test-piece begins to stretch rapidly, the extensometer is removed—rapid extension is a sign that fracture is imminent, and failure to remove the extensometer from the test-piece would probably lead to the destruction of the extensometer. The maximum force applied to the test-piece before fracture is measured.

4.21 A force/extension diagram can then be plotted (fig. 4.1). At first, the amount of extension is very small, compared with the increase in force. Such extension as there is is directly proportional to the force; that is, OA is a straight line. If the force is released at any point before A is reached, the test-piece will return to its original length. Thus the extension between O and A is *elastic*, and the material obeys Hooke's Law:

$$\text{stress} \propto \text{strain}$$

or

$$\frac{\text{stress}}{\text{strain}} = \text{a constant } (E)$$

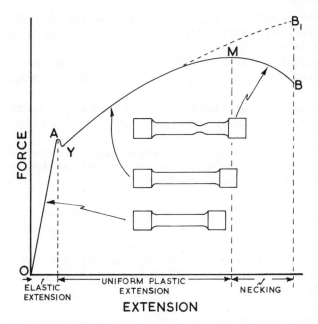

Fig. 4.1—The force-extension diagram for an annealed low-carbon steel.

This constant, *E*, is known as Young's Modulus of Elasticity for the material.

4.22 If the test-piece is stressed past the point *A* (known as the *elastic limit* or *limit of proportionality*), the material suddenly 'gives'; that is suffers a sudden extension for very little increase in force. This is called the *yield point* (*Y*), and, if the force is now removed, a small permanent extension will remain in the material. Any extension which occurs past the point *A* is of a *plastic* nature.

4.23 As the force is increased further, the material stretches rapidly—first uniformly along its entire length, and then locally to form a 'neck'. This 'necking' occurs just after the maximum force has been reached, at *M*, and since the cross-section decreases rapidly at the neck, the force at *B* required to break the specimen is *much less than* the maximum load at *M*.

This might be an appropriate moment to point out the difference between a force/extension diagram and a stress/strain diagram, since the terms are often loosely and incorrectly used. Figure 4.1 clearly represents a force/extension diagram, since the total force is plotted against the total extension, and, as the force decreases past the point *M*, for the reasons mentioned above, this decrease is indicated by the diagram. If, however, we wished to plot *stress* (force applied *per unit area* of cross-section of the specimen), we would need to measure the minimum diameter of the specimen, as well as its length, *for each increment of force*. This would be particularly important for values of force after the point *M*, since in this part of the test the diameter is decreasing rapidly, due to the formation of the 'neck'. Just as a chain is only as strong as its weakest link, so the test-piece is only as strong as the force its minimum diameter will support.

4.24 Thus, if stress were calculated on this decreasing diameter, the resulting stress/strain diagram would follow a path as indicated by the broken line to B_1.

In practice, however, a *nominal* value of the *tensile strength* of a material is calculated, using the maximum force (at *M*) and the original cross-sectional area of the test-piece. Thus,

$$\text{tensile strength} = \frac{\text{maximum force used}}{\text{original area of cross-section}}$$

The term '*engineering* stress' is often used; it signifies the force at any stage of the loading cycle divided by the *original* area of cross-section of the material.

4.25 Whilst tensile strength is a useful guide to the mechanical properties of a material, it is not of paramount importance in engineering design. After all, the engineer is not particularly interested in the material once plastic flow begins—unless he happens to be a production engineer interested in deep-drawing, or some other forming process. In terms of structural or constructional engineering, the elastic limit, *A*, will be of far greater significance.

4.26 The force/extension diagram shown in fig. 4.1 is typical of a low-carbon steel in the normalised condition. Unfortunately, force/extension diagrams for heat-treated steels, and for most other alloys, do not often show a well-defined yield point, and the 'elastic portion' of the graph merges gradually into the 'plastic section', as shown in the examples in fig. 4.2.

This makes it almost impossible to assess the yield stress of such an alloy, and, in cases like this, yield stress is replaced by a value known as *proof stress*. Thus the 0·1% proof stress of an alloy (denoted by the symbol $R_{p0·1}$) is that stress which will produce a permanent extension of 0·1% in the gauge length of the test-piece. This is very roughly equivalent to the permanent extension remaining in a normalised steel at its yield point.

4.27 The 0·1% proof stress of a material is derived as shown in fig. 4.3. The relevant part of the force/extension diagram is plotted as described earlier. A distance *OA*, equal to 0·1% of the gauge length, is marked along

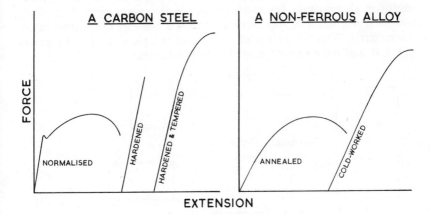

Fig. 4.2—Typical force-extension diagrams for both carbon steels and non-ferrous materials.

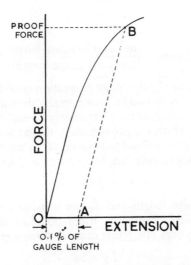

Fig. 4.3—The determination of 0·1% proof stress.

the horizontal axis. A line is then drawn from *A*, parallel to the straight-line portion of the force/extension diagram. The line from *A* intersects the diagram at *B*, and this indicates the proof force which would produce a permanent extension of 0·1% in the gauge length of the specimen. From this value of force, the 0·1% proof stress can be calculated.

$$0·1\% \text{ proof stress} = \frac{\text{proof force}}{\text{original cross-sectional area of test-piece}}$$

4.28 In addition to determining the tensile strength and the 0·1 % proof stress (or, alternatively, the yield stress), the percentage elongation of the test-piece at fracture is also derived. This is a measure of the ductility of the material. The two halves of the broken test-piece are fitted together (fig. 4.4), and the extended gauge length is measured.

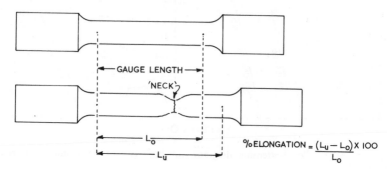

$$\% \text{ELONGATION} = \frac{(L_u - L_o)}{L_o} \times 100$$

Fig. 4.4—The determination of elongation (%).

$$\% \text{ elongation} = \frac{\text{increase in gauge length}}{\text{original gauge length}} \times 100$$

In order that values of percentage elongation shall be comparable, it is obvious that test-pieces should be geometrically similar; that is, there must be a standard relationship or ratio between cross-sectional area and gauge length. Test-pieces which are geometrically similar and fulfil these conditions are known as *proportional* test-pieces. They are generally circular in cross-section. BSI lays down that, for proportional test-pieces,

$$L_o = 5{\cdot}65\sqrt{S_o}$$

where L_o is the gauge length and S_o the original area of cross-section. This formula has been accepted by international agreement, and SI units are used. For test-pieces of circular cross-section, it gives a value

$$L_o = 5d$$

where d is the diameter at the gauge length. Thus a test-piece 200 mm² in cross-sectional area will have a diameter of 15·96 mm (16 mm) and a gauge length of 80 mm.

4.29 Tensile-testing machines vary both in design and capacity. Large machines applying forces of up to 1 MN or more are in use; whilst, at the other end of the scale, the miniature Hounsfield tensometer (fig. 4.5), with a capacity of 20 kN, is a portable bench instrument in which the tensile force is applied by means of a spring beam. This method has the particular

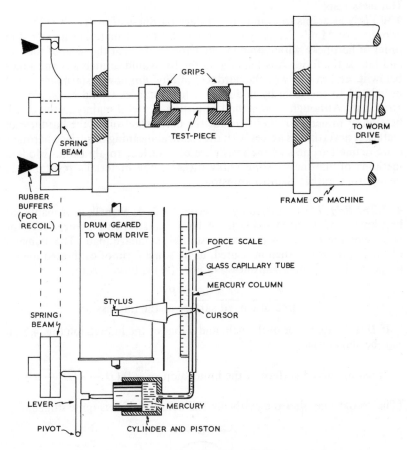

Fig. 4.5—The principles of the Hounsfield tensometer.
The upper part of the diagram shows the method of using the spring beam to apply a force to the test-piece. The lower part shows how the deflection of the beam is magnified, and measured by means of a mercury column.

advantage that, when necking of the specimen begins, the force applied by the spring beam is automatically reduced, and this enables the operator to plot the complete force/extension diagram. This is not possible with many large machines in which there is no provision for reducing the force once necking begins. With the Hounsfield tensometer, a force/extension diagram is plotted on special graph paper attached to a rotating drum which is geared to the actual extension of the test-piece. The operator follows the level of the mercury column (which records the force on the spring beam) by using the cursor. At appropriate intervals, he 'pricks' the graph paper with the stylus, so that a force/extension diagram is traced out as a series of pin-pricks.

Hardness tests

4.30 Early attempts to evalute the surface hardness of materials led to the adoption of Moh's Scale. This consists of a list of materials arranged in order of hardness, with diamond, the hardest of all, at the head of the list, and talc at the foot. Any material in the list would scratch any substance below it, and in this way the hardness of any 'unknown' substance could be related to the scale by finding which substances would or would not scratch it. Although this method of testing is useful mainly in the classification of minerals, rather than for the determination of the hardness of metals, it nevertheless agrees with the classical meaning of surface hardness if we define hardness as the resistance of a surface to abrasion. Modern methods of hardness-testing really measure the material's resistance to penetration, rather than to abrasion.

4.31 *The Brinell test,* devised by a Swede, Dr J. A. Brinell, is probably the best known of the hardness tests. A hardened steel ball is forced into the surface of a test-piece by means of a suitable standard load. The diameter of the impression is then measured, using some form of calibrated microscope, and the Brinell Hardness Number (H) is found from:

$$H = \frac{\text{load } (P)}{\text{area of curved surface of the impression}}$$

If D is the diameter of the ball, and d that of the impression (fig. 4.6), it can be shown that:

$$\text{area of curved surface of the impression} = \frac{\pi}{2} D (D - \sqrt{D^2 - d^2})$$

(The reader may find the mathematical solution of this quite difficult.)

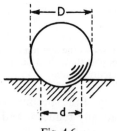

Fig. 4.6.

It follows that
$$H = \frac{P}{\frac{\pi}{2} D (D - \sqrt{D^2 - d^2})}$$

To make tedious calculations unnecessary, H is generally found by reference to tables which relate H to d—a different set of tables being used for each possible combination of P and D.

In carrying out a Brinell test, certain conditions must be fulfilled. First, the depth of impression must not be too great relative to the thickness of the test-piece, otherwise we may produce the situation shown in fig. 4.7A.

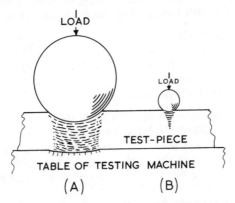

Fig. 4.7—This illustrates the necessity of using the correct ball diameter in relation to the thickness of the test-piece.

Here it is the table of the machine, rather than the test-piece, which is supporting the load. For soft materials, the thickness of the test-piece should be at least fifteen times the *depth* of the impression; whilst for hard materials it should be at least seven times the depth of the impression.

4.32 Balls of 10 mm, 5 mm, and 1 mm diameter are available; so one appropriate to the thickness of the test-piece should be chosen, bearing in mind that the larger ball it is possible to use, the more accurate is the result likely to be. Having decided upon a suitable ball, we must now select a load which will produce an impression of reasonable proportions. If, for example, in testing a soft metal we use a load which is too great relative to the size of the ball, we shall get an impression similar to that indicated in fig. 4.8 (i). Here the ball has sunk to its full diameter, and the result is meaningless. On the other hand, the impression shown in fig. 4.8 (ii)

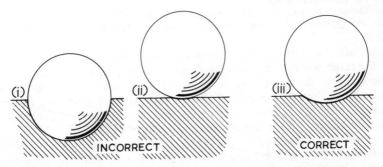

Fig. 4.8—It is essential to use the correct P/D^2 ratio for the material being tested.

would be obtained if the load were too small relative to the ball diameter, and here the result would be uncertain. For different materials, then, the ratio P/D^2 has been standardised in order to obtain accurate and comparable results. P is still measured in 'kg force' and D in mm.

Material	P/D^2
steel	30
copper alloys	10
aluminium alloys	5
lead alloys and tin alloys	1

As an example, in testing a piece of steel, we can use a 10 mm ball in conjunction with a 3000 kgf load, a 5 mm ball with a 750 kgf load, or a 1 mm ball with a 30 kgf load. As mentioned above, the choice of ball diameter (D) will rest with the thickness of the test-piece; whilst the load to be used with it will be determined from the appropriate P/D^2 ratio.

4.33 *The Vickers pyramid hardness test* uses a square-based diamond pyramid (fig. 4.9) as the indentor. One great advantage of this is that all

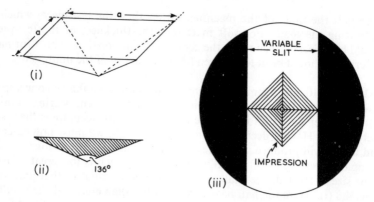

Fig. 4.9—The Vickers pyramid hardness test.
(i) The diamond indentor.
(ii) The angle between opposite faces of the diamond is 136°.
(iii) The appearance of the impression in the microscope eye-piece.

impressions will be geometrically similar, and, within limits, the accuracy of the result will not vary with the depth of the impression. Consequently, the operator does not have to choose a P/D^2 ratio as he does in the Brinell test, though he must still observe the relationship between depth of impression and thickness of specimen, for reasons similar to those indicated in the case of the Brinell test, and illustrated in fig. 4.7.

A further advantage of the Vickers hardness test is that hardness values for very hard materials (above an index of 500) are likely to be more accurate than the corresponding Brinell numbers—a diamond does not deform under high pressure to the same extent as does a steel ball, and so the result will be less uncertain.

4.34 In this test, the diagonal length of the square impression is measured by means of a microscope which has a variable slit built into the eyepiece (fig. 4.9 (iii)). The width of the slit is adjusted so that its edges coincide with the corners of the impression, and the relative diagonal length of the impression is then obtained from a small instrument geared to the movement of the slit, and working on the principle of a revolution counter. The ocular reading thus obtained is converted to Vickers Pyramid Hardness Number (VPN) by reference to tables. The size of the impression is related to hardness in the same way as is the Brinell Number.

In some quarters this test is referred to as the Diamond Pyramid Hardness (DPH) test.

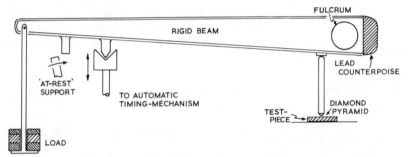

Fig. 4.10—The loading system for the Vickers pyramid hardness machine.
It is essentially a second-order lever system. The fifteen-second period of load application is timed by an oil dashpot system.

4.35 Since the impression made by the diamond is generally much smaller than that produced by the Brinell indentor, a smoother surface finish is required on the test-piece. This is produced by rubbing with fine emery paper of about '400 grit'. At the same time, surface damage is negligible, making the Vickers test more suitable for testing finished components.

The specified time of contact between the indentor and the test-piece in most hardness tests is 15 seconds. In the Vickers testing machine, this period of contact is timed automatically by a piston working in an oil dashpot.

4.36 *The Rockwell test* was devised in the USA. It is particularly useful for rapid routine testing of finished material, since the hardness number is indicated directly on a dial, and no subsequent measurement of the diameter of the impression is involved. Although the depth (h) (fig. 4.11) of the impression is measured by the instrument, this is converted (on the dial) to hardness values in which the surface area of the impression is related to the load in the usual way.

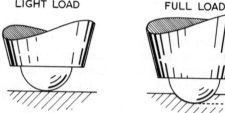

Fig. 4.11—The Rockwell test.

The test-piece, which needs no preparation save the removal of dirt and scale from the surface, is placed on the table of the instrument, and the indentor is brought into contact with the surface under 'light load'. This takes up the 'slack' in the system, and the scale is then adjusted to zero. 'Full load' is then applied, and when it is subsequently released (timing being automatic), the test-piece remains under 'light load' whilst the hardness index is read direct from the scale.

4.37 There are several different scales on the dial, the most important of which are:

(1) scale B, which is used in conjunction with a $\frac{1}{16}''$ diameter steel ball and a 100 kgf load;

(2) scale C, which is used with a diamond cone of 120° angle and a 150 kgf load;

(3) scale A, which is used in conjunction with the diamond cone and a 60 kgf load.

Of the scales available, possibly the most useful are scale C, which is used mainly for hardened steels and other very hard materials; and scale B, which is used for most other materials, including normalised steels and non-ferrous alloys.

The Rockwell machine is very rapid in action, and can be used by relatively unskilled operators. Since the size of the impression is also very small, it is particularly useful for the routine testing of stock or individual components on a production basis.

4.38 *The Shore scleroscope* (Greek *skleros* meaning 'hard') is a small portable instrument which can be used for testing the hardness of large components such as rolls, drop-forging dies, castings, and gears. Such components could not be placed on the table of one of the more orthodox machines mentioned above. The scleroscope embodies a small diamond-tipped 'tup', or hammer, of mass approximately 2·5 g, which is released so that it falls from a standard height of about 250 mm inside a graduated glass tube placed on the test surface. The height of rebound is taken as the hardness index.

4.39 Soft materials absorb more of the kinetic energy of the hammer, as they are more easily penetrated by the diamond point, and so the height of rebound is less. Conversely, a greater height of rebound is obtained from hard materials.

Impact tests
4.40 These tests are used to indicate the toughness of a material, and particularly its capacity for resisting mechanical shock. Brittleness, resulting from a variety of causes, is often *not* revealed during a tensile test. For example, nickel-chromium constructional steels suffer from a

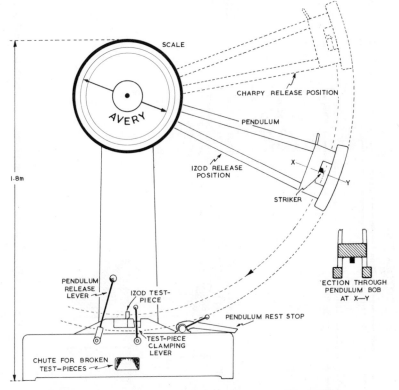

Fig. 4.12—The Avery-Denison Universal impact-testing machine.
This machine can be used for either Charpy or Izod impact tests. For Izod tests, the pendulum is released from the lower position, to give a striking energy of 170 J; and for the Charpy test it is released from the upper position, to give a striking energy of 300 J. (The scale carries a set of graduations for each test.) The machine can also be used for impact-tension tests.

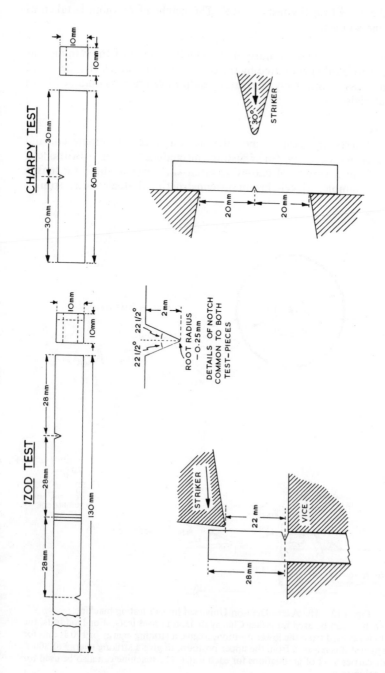

Fig. 4.13—Details of standard test-pieces used in both the Izod and Charpy tests.

defect known as 'temper brittleness' (13.23). This is caused by faulty heat-treatment, yet a tensile test-piece derived from satisfactorily treated material and one produced from similar material but which had been incorrectly heat-treated might both show approximately the same tensile strengths and elongations. In an impact test, however, the unsatisfactory material would prove to be extremely brittle as compared with the correctly treated one, which would be tough.

4.41 *The Izod impact test* employs a standard notched test-piece (fig. 4.13) which is clamped firmly in a vice. A heavy pendulum, mounted on ball-bearings, is allowed to strike the test-piece after being released from a fixed, height (fig. 4.12). The striking energy of approximately 163 J is partially absorbed in breaking the test-piece, and as the pendulum swings past, it carries with it a drag pointer which it leaves at its highest point of swing. This indicates the amount of mechanical energy used in fracturing the test-piece. To set up stress concentrations which will ensure that fracture does occur, the test-piece is notched. It is essential, however, that this notch always be standard, for which reason a standard gauge is supplied to test the dimensional accuracy of the notch, both in this and the other impact tests dealt with below.

4.42 *The Charpy impact test* is of continental origin, and differs from the Izod test in that the test-piece is supported at each end (fig. 4.13); whereas the Izod test uses a test-piece held cantilever fashion. Here the load on the pendulum can be varied so that the impact energy is either 150 J or 300 J.

4.43 *The Hounsfield balanced impact test* makes use of a portable machine in which the striking energy is produced by means of two pendulums which move in opposite directions to each other. One pendulum is U-shaped (fig. 4.14), so that the other may swing through the gap provided. The test-piece, mounted in the inner pendulum, is struck at each of its ends by the outer pendulum as the two pendulums swing past each other. The net effect is similar to that obtained in the Charpy test. A drag pointer moves over a calibrated scale, as in the tests described above, and similarly indicates the energy used in breaking the test-piece under conditions of shock loading.

4.44 The total mass of the apparatus is of the order of only 18 kg, yet it produces a striking energy of some 65 J, because the pendulums move in opposition. More important still, nearly all of the energy absorbed is dissipated in the moving test-piece, and little shock is transmitted; whereas in both the Izod and Charpy machines, much energy is absorbed by the rigid vice, which must be set in massive foundations for this reason. A similar principle is used in modern high-energy rate-forging, where *both* halves of the die move to meet at the work-piece, thus eliminating the need for a massively mounted anvil.

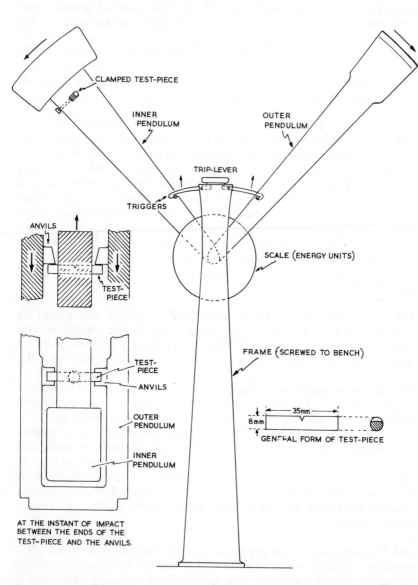

CLAMPED TEST-PIECE

INNER PENDULUM

OUTER PENDULUM

TRIP-LEVER

TRIGGERS

ANVILS

TEST-PIECE

SCALE (ENERGY UNITS)

FRAME (SCREWED TO BENCH)

TEST-PIECE

ANVILS

OUTER PENDULUM

INNER PENDULUM

AT THE INSTANT OF IMPACT
BETWEEN THE ENDS OF THE
TEST-PIECE AND THE ANVILS.

35mm

8 mm

GENERAL FORM OF TEST-PIECE

Fig. 4.14—The principles of the Hounsfield balanced impact-testing machine.

Creep

4.50 When stressed over a long period of time, some metals extend very gradually, and may ultimately fail at a stress *well below* the tensile strength of the material. This phenomenon of slow but continuous extension under a steady force is known as 'creep'. Such slow extension is more prevalent at high temperatures, and for this reason the effects of creep must be taken into account in the design of steam and chemical plant, gas and steam turbines, and furnace equipment.

4.51 The 'limiting creep stress' of a material at any given temperature is the maximum stress it can withstand without showing any measurable extension. Creep tests are carried out on test-pieces which are similar in form to ordinary tensile test-pieces. A test-piece is enclosed in a thermo-statically controlled electric tube furnace which can be maintained accurately at a fixed temperature over the long period of time occupied by the test. The test-piece is statically stressed, and some form of sensitive extensometer is used to measure the extremely small extension at suitable time intervals. A set of creep curves, obtained for different static forces at the same temperature, is finally produced, and from these the limiting creep stress is derived.

4.52 Since we are dealing with such very small extensions, it is difficult to estimate the stress which just fails to produce any 'measurable extension'. We are, in fact, limited by the sensitivity of the extensometer, and it is often more practical to determine the stress which will produce a definite amount of creep in a fixed time—say two or three days. Such a value is generally referred to as the 'time-yield stress' for those conditions.

Fatigue

4.60 Whenever failure of some structure or machine leads to a disaster which is worthwhile reporting, the press invariably hints darkly at 'metal fatigue'. Unfortunately, the technical knowledge of these gentlemen of the press is often suspect, as is evidenced by that oft-repeated account of the legendary cable 'through which a current of 132 000 volts was flowing'.

4.61 Although the public was first made aware of the phenomenon of fatigue following investigations of the Comet airliner disasters some years ago, the underlying principles of metal fatigue were appreciated more than a century ago by the British engineer Sir William Fairbairn, who carried out his classical experiments with wrought-iron girders. He found that a girder which would support a static load of 12 tonf for an indefinite period would nevertheless *fail* if a load of only 3 tonf were raised and lowered on it some three million times. Fatigue is associated with the effects which a fluctuating or an alternating force may have on a member.

4.62 Following the work of Fairbairn, the German engineer Wöhler, with native inventiveness, produced the well known fatigue-testing machine which still bears his name. This is a device (fig. 4.15) whereby alternations

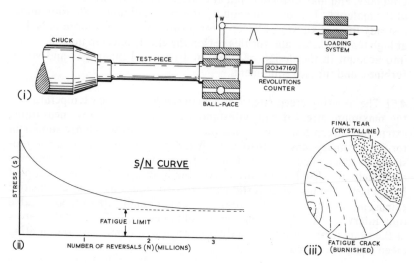

(i)

S/N CURVE

FATIGUE LIMIT

(ii) NUMBER OF REVERSALS (N)(MILLIONS)

FINAL TEAR (CRYSTALLINE)

(iii) FATIGUE CRACK (BURNISHED)

Fig. 4.15—(i) The principle of a simple fatigue-testing machine.
(ii) A typical *S/N* curve obtained from a series of tests.
(iii) The appearance of the fractured surface of a shaft which has failed due to fatigue.

of stress can be produced in a test-piece very rapidly, and so reduce to a reasonable period the time required for a fatigue test. As the test-piece turns through 180°, the force W acting at a point on the specimen falls to zero, and then increases to W in the opposite direction. To find the fatigue limit, a number of similar specimens of the material are tested in this way, each at a different value of W, until failure occurs, or, alternatively, until about 20 million reversals have been endured. (It is, of course, not possible to subject the test-piece to the ideal infinite number of reversals.) From these results, an *S/N* curve is plotted; that is, stress (*S*) against the number of reversals (*N*) endured (fig. 4.15 (ii)). The curve becomes horizontal at a stress which will be endured for an infinite number of reversals. This stress is the *fatigue limit* or *endurance limit*. Some non-ferrous materials do not show a well defined fatigue limit; that is, the *S/N* curve slopes gradually down to the horizontal axis.

4.63 A fatigue fracture has a characteristic type of surface, and consists of two parts (fig. 4.15 (iii)). One is smooth and burnished, and shows ripple-like marks radiating outwards from the centre of crack formation; whilst the other is coarse and crystalline, indicating the final fracture of the remainder of the cross-sectional area which could no longer withstand the load.

4.64 Fatigue failure will ultimately occur in any member which is stressed above its fatigue limit in such a way that the operating stress fluctuates or alternates. Such failure can be due simply to bad design and lack of understanding of fatigue, but is much more likely to be due to the presence of unforeseen high-frequency vibrations in a member which is stressed above the fatigue limit. This is possible since the fatigue limit is well below the tensile strength for all materials.

Some time ago, the author's opinion was sought on the possible cause of failure of a short length of copper tube which had held, cantilever fashion, a small pressure gauge, in the manner shown in fig. 4.16. The broken tube

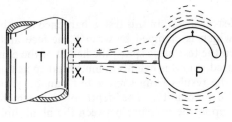

Fig. 4.16.

was sent with the information that it had broken off at XX_1 under the action of no other force than the weight of the pressure gauge (P), which was very small. The fracture appeared, on examination, to be of a typical fatigue nature. This tentative suggestion brought forth derisory comments from the engineer in charge, who claimed that the copper tube must be of poor quality, since failure had taken place after only a few days of service in a *static* condition. On ultimately visiting the site, the author's suspicions were soon confirmed—the pipe (T) was connected to a large air-compressor, the nerve-shattering vibrations from which were apparent at a distance of several hundred metres.

4.65 Any feature which increases stress concentrations may precipitate fatigue failure. Thus a fatigue crack may start from a keyway, a sharp fillet, a microstructural defect, or even a bad tool mark on the surface of a component which has otherwise been correctly designed with regard to the fatigue limit of the material from which it was made.

4.70 Other mechanical tests

4.71 *The Erichsen cupping test*. Materials used for deep-drawing are inevitably those of very high ductility. However, a simple measurement of ductility in terms of percentage elongation (obtained during the tensile test) does not always give a complete assessment of deep-drawing properties. A test which imitates the conditions present during a deep-drawing operation is often preferable. Such a test is the Erichsen cupping test, and it is commonly used to assess the deep-drawing quality of soft brass, aluminium,

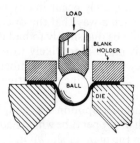

Fig. 4.17—The principle of the Erichsen cupping machine.

copper, or mild-steel sheet. In this test, a hardened steel ball is forced into the test-piece, which is clamped between a die-face and a blank-holder (fig. 4.17). When the test-piece splits, the height of the cup which has been formed is measured, and this height (in mm) is taken as the Erichsen value. Unfortunately, the results from such a test can be variable, even with material of uniform quality. The depth of cup which can be drawn depends largely upon the pressure between the blank-holder and the die-face. Light pressure will allow metal to be drawn between the die and holder, and a deeper cup is formed. The continentals tend to use firm pressure between die and holder; whereas in Britain we 'cheat' by using lighter pressure. Nevertheless the British Standards Institution has now recognised the Erichsen test in BS 3855 where test piece dimensions, testing apparatus, lubricants and precautions are dealt with.

4.72 Possibly the most useful aspect of the test is that it gives some idea of the grain size of the material, and hence its suitability for deep-drawing, as indicated in fig. 4.18.

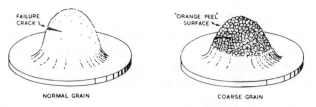

Fig. 4.18—Erichsen test-pieces.

4.73 *Bend tests* are often used as a means of judging the suitability of a metal for similar treatment during a production process. For example, copper, brass, or bronze strip used for the manufacture of electrical switch-gear contacts by simple bending processes may be tested by a bending operation which is somewhat more severe than that which will be experienced during production.

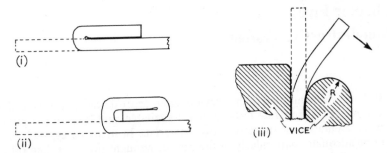

Fig. 4.19—Simple bend tests.
(i) The material is merely bent back upon itself.
(ii) Here it is doubled over its own thickness, the second bend being the test bend.
(iii) Here a specific radius is used.

Figures 4.19 (i) and (ii) illustrate simple tests requiring little equipment other than a vice; whilst (iii) suggests a test where the material is bent round some specific radius (R).

The surface affected by the bending process is examined for cracks, and, if necessary, for coarse grain ('orange peel').

4.74 *Compression tests* are used mainly in connection with cast iron, since this is a material more likely to be used under the action of compressive forces than in tension. A cylindrical block, the length of which is twice its diameter, is used as a test-piece. This is compressed (using a tensile-testing machine running in 'reverse') until it fails.

Malleable metals do not show a well defined point of failure (fig. 4.20 (ii)),

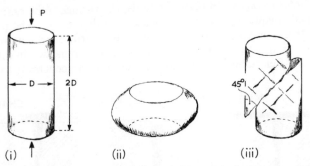

Fig. 4.20—The behaviour of brittle and ductile materials during a compression test.

but with brittle materials (iii), the *ultimate compressive stress* can be measured accurately, since the material fails suddenly, usually by multiple shear at angles of 45° to the direction of compression.

Chapter Five
Non-destructive Testing

5.10 The tests mentioned in the previous chapter are destructive tests, which are applicable if separate test-pieces are available, and if these test-pieces are representative of the production material. In many cases, such tests will be adequate, particularly in the case of wrought products, like rolled strip and drawn rod, which are generally very uniform in properties throughout a large batch of material. Parts which are produced individually, however, such as castings and welded joints, may vary widely in quality purely because they are made individually and under the influence of so many variable factors. If the quality of such components is important, and the expense thus justified, it may be necessary to test each component individually by some type of non-destructive test. Such tests seek to detect faults and flaws either at the surface or below it, and a number of suitable methods is available in each case.

The detection of surface cracks

5.20 Surface cracks may be produced in a component as a result of heat-treatment or, in a welded joint, by contraction during cooling. In some cases, such cracks may be discovered by careful visual inspection, with or without the aid of a hand magnifier. The presence of very fine hair-line cracks, however, is less easy to detect, and some aid is generally necessary.

5.21 *Penetrant methods.* In these methods, the surface to be examined is first cleaned adequately to remove grease, and then dried. The penetrant liquid is then sprayed or swabbed on to the surface, which should be warmed to about 90°C. Small components may be immersed in the heated penetrant. After sufficient time has elapsed for the penetrant to fill any

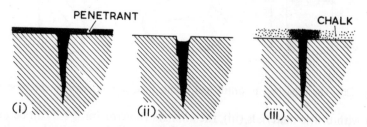

Fig. 5.1—The penetrant method of crack detection.
(i) The cleaned surface is coated with the penetrant, which seeps into any cracks.
(ii) Excess penetrant is cleaned from the surface.
(iii) The surface is coated with chalk. The penetrant is expelled from the crack by the contraction of the cooling metal, and stains the chalk.

cracks, the excess is carefully washed from the surface with warm water (the surface tension of water is too high to allow it to enter the fine cracks). The test surface is then carefully dried, coated with fine powdered chalk, and set aside for some time. As the coated surface cools, it contracts, and penetrant tends to be squeezed out of any cracks; so the chalk layer will become stained, thus revealing the presence of the cracks. Most penetrants of this type contain a scarlet dye which renders the stain immediately noticeable.

5.22 In some cases, the penetrant contains a compound which becomes fluorescent under ultra-violet light. The use of chalk is then unnecessary. When the prepared surface is illuminated by ultra-violet light, the cracks containing the penetrant are revealed as bright lines on a dark background.

Penetrant methods in general are particularly useful for the examination of non-ferrous metals and austenitic (non-magnetic) steels. Aluminium-alloy castings are frequently examined in this way.

5.23 *Magnetic dust methods.* These methods can be applied only to magnetic materials, but nevertheless provide a quick and efficient method of detecting cracks. A further advantage over the penetrant method is that cracks

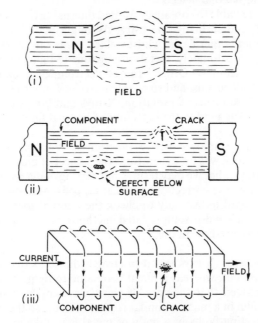

Fig. 5.2—Magnetic methods of crack detection.
(i) The principle of the process—lines of force tend to 'spread' in an air gap.
(ii) A magnetic 'circuit'. (iii) Using a high current to induce a field.

immediately *below* the surface are also detected; so the magnetic dust method is particularly suitable for examining machined or polished surfaces, where the mouth of a crack may well have become 'burred' over.

The magnetic method involves making the component part of a 'magnetic circuit' (fig. 5.2 (ii)), or, alternatively, passing through it a heavy current at low e.m.f., so that a magnetic field will be induced in the component (fig. 5.2 (iii)).

No matter which method is used to produce the magnetic field, the object is to 'saturate' the component with lines of force. These lines of force pass easily through a magnetic material, which is, so to speak, a 'good conductor' of lines of force. In air, they repel each other to the extent that they spread as indicated in fig. 5.2 (i). Consequently, on meeting a gap or other discontinuity at or just below the surface, the lines of force spread outwards (fig. 5.2 (ii) and (iii)). If some iron dust is now sprinkled on the surface of the component, it will stick to the surface where the lines of force break out, thus revealing the site of the fault. Alternatively, the magnetised component can be placed in paraffin containing a suspension of tiny iron particles. The particles will be attracted to the surface of the component at any points where, due to the presence of a fault, lines of force cut through.

The detection of internal defects

5.30 Castings are liable to contain unwanted internal cavities, in the form of gas blow-holes and shrinkage porosity. Wrought materials may contain slag inclusions and other flaws; whilst welded joints can suffer from any of these defects. Unfortunately, metals are opaque to light; so these internal faults are hidden from us. Other forms of electromagnetic radiation, however, will penetrate metals, and so enable us to 'see' into the interior of the material. Of these forms of radiation, X-rays and γ-rays ('gamma' rays) are the most widely used in the detection of internal faults.

5.31 The railway employee whose function it is to tap the wheels of railway rolling-stock was for generations the subject of music-hall humour. In fact he employs what we may term 'sonic testing'—if a wheel is cracked, it no longer emits the correct 'ring' when tapped. Methods such as this have also been used industrially to check the internal soundness of components, but much more sophisticated methods of testing, using 'ultrasonic' vibrations, are now in use.

5.32 *X-ray methods.* X-rays travel in straight lines, in the same manner as do light rays; but, whereas a metal is opaque to light, it is transparent to X-rays, provided that they are of short wavelength. X-rays affect a photographic film in a manner similar to that of light; so the most efficient method of detecting faults in a body of metal is to take an X-ray photograph of its interior. The reader may be familiar with a similar application of X-rays in medicine. There is a difference in the nature of the X-rays used, however, since those employed in the radiography of metals need to be 'harder', that is, of shorter wavelength, so that they will penetrate metals more effectively. These hard X-rays would damage our body tissues, and for this reason the equipment must be well shielded to prevent stray radiations from reaching the operator. A method of shielding is indicated in fig. 5.3. Here the X-ray tube is emitting radiation which passes through

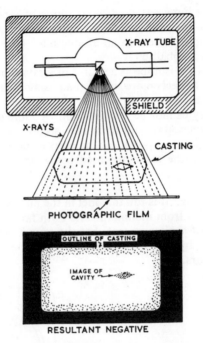

Fig. 5.3—Radiography of a casting, using X-rays.

the casting, and forms an image on a photographic film placed behind the casting. X-rays will penetrate that region of the casting containing the cavity much more easily, and so produce a *denser* image on the film. (Readers who have an interest in photography will know that those areas on a monochrome negative which receive the most light are densest and darkest when the negative has been developed.)

5.33 A fluorescent screen may be substituted for the photographic film, so that the resultant radiograph may be viewed instantaneously. This type of fluoroscopy is much cheaper and quicker, but is less sensitive than photography, and its use is limited to the less-dense metals.

5.34 *γ-ray methods.* γ-rays can also be used in the radiography of metals. They are 'harder' than X-rays, and are therefore able to penetrate a greater thickness of metal or, alternatively, a more dense metal. Hence they are particularly useful in the radiography of steel, which absorbs radiation more effectively than do the light alloys. As might be expected, exposure to γ-radiation is extremely dangerous. It is in fact the lethal radiation emitted from the notorious strontium-90, present in the fall-out products of an atomic-bomb explosion.

5.35 Originally, radium was used as a source of γ-rays, but artificially activated isotopes are now generally employed. These are prepared by irradiating a suitable element in an atomic pile. One of the most useful activated isotopes is cobalt-60.

Manipulation of the isotope as a source of γ-rays in metallurgical radiography is in many respects more simple than using X-rays, though security arrangements are even more important. Not only are γ-rays harder, but, unlike X-rays, they cannot be switched off. A radioactive isotope emits continuously for a period varying from a few seconds to thousands of years, depending upon its type. γ-rays can be used to radiograph considerable thicknesses of steel, the technique used being similar to that used with X-rays.

5.36 *Ultrasonic testing.* It is fun to yodel in the mountains, and listen to a succession of echoes from mountain-sides both far and near. Ultrasonic testing is somewhat similar, except that the sound waves generated by a yodel are replaced by very high-frequency vibrations, which are beyond the range of our ears.

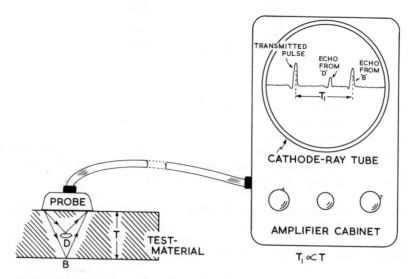

Fig. 5.4—The use of ultrasonic vibrations in detecting flaws below the surface in metal plate.

Figure 5.4 represents the principles of ultrasonic testing. A probe containing a quartz crystal, which can both transmit and receive high-frequency vibrations, is passed over the surface of the material to be tested. The probe is connected to an amplifier, which converts and amplifies the signal, before it is recorded on the cathode-ray tube.

5.37 Under normal conditions, the vibrations will pass from the probe, unimpeded through the metal, and be reflected from the bottom inside surface at B back to the probe, which also acts as receiver. Both the transmitted pulse and its echo are recorded on the cathode-ray tube, and the distance, T_1, between the 'blips' is proportional to the thickness, T, of the test material. If any discontinuity is encountered, such as the blow-hole, D, then the pulse is interrupted, and reflected back as indicated. Since this echo returns to the receiver in a shorter time, an intermediate 'blip' appears on the cathode-ray tube. Its position relative to the other 'blips' indicates the distance of the fault below the surface. Different types of probe are available for materials of different thickness.

Chapter Six
Mechanical Deformation and Recrystallisation of Metals

6.10 When a piece of metal is shaped by the application of stress, deformation takes place in two stages, as indicated during a tensile test (4.22). At first, crystals within the metal are distorted in an elastic manner, and this distortion increases proportionally with the increase in stress. If the stress is removed during this stage, the metal returns to its original shape, illustrating the elastic nature of the deformation. On the other hand, if the stress is increased further, a point is reached where the forces which bind together the atoms in the lattice structure are overcome to the extent that layers or planes of atoms begin to slide over each other. This process of 'slip', as metallurgists call it, is *not* reversible; so, if the stress is now removed, permanent deformation remains in the metal (fig. 6.1). This type of deformation is termed 'plastic'.

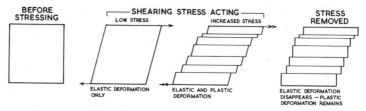

Fig. 6.1—Deformation of a metal by both elastic and plastic means.

6.11 Microscopic evidence of the nature of slip is fairly easily obtained. If a piece of pure copper or aluminium is polished and etched, and then squeezed in a vice with the polished face uppermost, so that it is not damaged, examination under the microscope shows a large number of parallel, hair-like lines or *slip bands* on the surface. These lines are shadows cast by minute ridges formed as blocks of atoms within each crystal slid over each other (fig. 6.2).

6.12 Slip of this type can occur along a suitable plane, until it is prevented by some fault or obstacle within the crystal. A further increase in the stress will then produce slip on another plane or planes, and this process goes on until all available slip planes in the piece of metal are used up. The metal is then said to be work-hardened, and any further increase in stress will lead to fracture. Microscopic examination will show that individual crystals have become elongated and distorted in the direction in which the metal was deformed. In this condition, the metal is hard and strong; but it has lost its ductility, and, if further shaping is required, it must be softened by annealing.

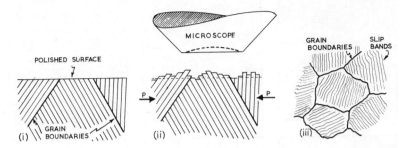

Fig. 6.2—The formation of slip bands.
(i) indicates the surface of the specimen before straining, and (ii) the surface after straining. The relative slipping along the crystal planes produces ridges, which cast shadows on the surface of the specimen when this is viewed through the microscope (iii).

Annealing and recrystallisation

6.20 A cold-worked metal, that is, one which has been deformed without any application of heat, is in a state of considerable internal stress. During annealing, these stresses are removed, and the original ductility of the metal returns. The changes which accompany an annealing process occur in three stages.

6.21 *The relief of stress.* As the temperature of the cold-worked metal is gradually raised, some of the internal stresses disappear, as atoms move through small distances into positions nearer equilibrium. At this stage, there is no alteration in the generally distorted appearance of the structure, and, indeed, the strength and hardness produced by cold-working remain high. Nevertheless, some hard-drawn materials, such as 70–30 brass, are given a low-temperature anneal in order to relieve internal stresses, as this reduces their tendency towards 'season cracking' during service, that is, the opening up of cracks along grain boundaries, due to the combined effects of internal stresses and surface corrosion.

6.22 *Recrystallisation.* With further increase in temperature, a point is reached where new crystals begin to grow from nuclei which form within the structure of the existing distorted crystals. These nuclei are generally produced where internal stress is greatest, that is, at grain boundaries and on slip planes. As the new crystals grow they take up atoms from the old distorted crystals which they gradually replace. Unlike the old crystals, which had become elongated in one direction by the cold-working process, these new crystals are small and equi-axed.

This phenomenon, known as *recrystallisation*, is the principal method used to obtain a tough, fine-grained structure in most non-ferrous metals. The minimum temperature at which it will occur is called the 'recrystallisation temperature' for that metal, though it is not possible to determine this

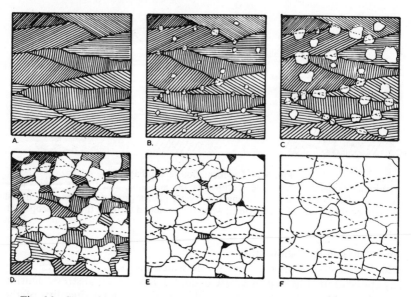

Fig. 6.3—Stages in the recrystallisation of a metal during an annealing process. (A) represents the metal in the cold-rolled state. At (B) recrystallisation has just begun, with the formation of new crystal nuclei. These grow by absorbing the old crystals, until at (F) recrystallisation is complete.

temperature precisely, because it varies with the amount of cold-work to which the metal had been subjected before the annealing process. The more heavily the metal is cold-worked, the greater the internal stress, and the lower the temperature at which recrystallisation will begin. Alloying, or the presence of impurities, raises the recrystallisation temperature of a metal.

Metal	Recrystallisation temperature (°C)
tungsten	1200
nickel	600
pure iron	450
copper	190
aluminium	150
zinc	20
lead ⎱ tin ⎰	below room-temperature; hence they cannot be 'cold-worked'

Table 6.1—Recrystallisation temperatures of some metals.

6.23 *Grain growth.* If the annealing temperature is well above that for recrystallisation of the metal, the new crystals will increase in size by absorbing each other cannibal-fashion, until the resultant structure becomes relatively coarse-grained. This is undesirable, since a coarse-grained material is generally less ductile than a fine-grained material of similar composition. Moreover, if the material is destined for deep-drawing, coarse grain tends to disfigure a stretched surface by giving it a rough, rumpled appearance known as 'orange peel'. Both the time and temperature of annealing must be controlled, in order to limit grain growth; though, as indicated by fig. 6.4, temperature has a much greater influence than does time.

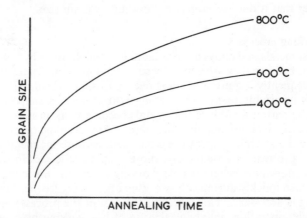

Fig. 6.4—The effects of time and temperature during annealing on the grain size of a previously cold-worked metal.

6.24 The amount of cold-work the material receives prior to being annealed also affects the grain size produced. In heavily cold-worked metal, a large number of nuclei will form at the recrystallisation temperature during annealing, and so the crystals will be small. On the other hand, very light cold-work (the 'critical' amount indicated in fig. 6.5) gives rise

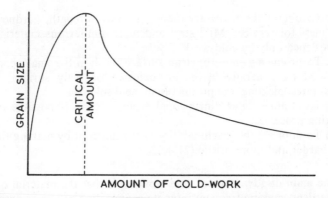

Fig. 6.5—The effect of the initial amount of cold-work on the grain size produced during annealing.

to few nuclei, because the metal is not highly stressed. Consequently the grain size will be large, and the ductility poor.

6.25 Alloy additions are often made to limit the grain growth of metals and alloys during heat-treatment processes. Thus, up to 5% nickel is added to case-hardening steels in order to reduce grain growth during the carburising process. When we say that nickel toughens a steel, we must remember that it does so largely by limiting its grain size.

Cold-working processes

6.30 Most metals and alloys are produced in wrought form by hot-working processes, because they are generally softer and more malleable when hot, and consequently require much less energy to shape them. Hot-working processes invariably make use of *compressive* forces to shape the work-piece. The reason for this is that metals become weak in tension at high temperatures; so their ductility decreases. Any attempt to pull a metal through a die at high temperature would fail, because the metal would be so weak as to tear. Consequently, those shaping processes in which tension is employed are generally cold-working processes. Since most metals work-harden quickly during cold-working operations, frequent inter-stage annealing is necessary. This increases the cost of the process; and, as far as possible, operations involving the use of hot-working by compression are used, rather than cold-working processes, which make frequent annealing stages necessary. Wire used to be made by drawing down a previously rolled bar. This involved many drawing operations, interspersed with annealing stages. Now, a cast billet is hot-extruded as thick wire in a single operation; this thick wire is then drawn down in a minimum of cold-working stages, through dies.

6.31 Other reasons for using cold-working processes are:

(1) To obtain the necessary combination of strength, hardness and toughness for service. Mild steel and most non-ferrous materials can be hardened only by cold-work.

(2) To produce a smooth, clean surface finish in the final operation. Hot-working generally leaves an oxidised or scaly surface, which necessitates 'pickling' the product in an acid solution.

(3) To attain greater dimensional accuracy than is possible in hot-working processes.

(4) To improve the machinability of the material by making the surface harder and more brittle (21.40).

6.32 The main disadvantage of cold-working is that the material quickly work-hardens, making frequent inter-stage annealing processes necessary. A typical sequence of operations for the manufacture of a cartridge-case

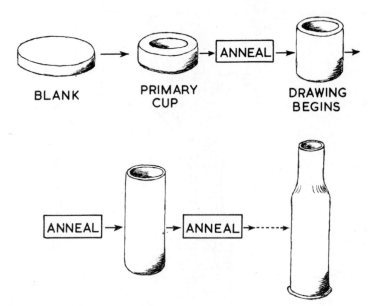

Fig. 6.6—Typical stages in the deep-drawing of a cartridge-case.
In practice, many more intermediate operations may be necessary, only the principle being indicated here.

by deep-drawing 70–30 brass sheet is shown in fig. 6.6. Here the inter-stage annealing processes will be carried out in an oxygen-free atmosphere ('bright annealing'), in order to make acid-pickling unnecessary.

6.33 Other cold-working processes include:

(1) the drawing of wire and tubes through dies;
(2) the cold-rolling of metal plate, sheet, and strip;
(3) spinning and flow-turning, as in the manufacture of aluminium kitchenware;
(4) stretch forming, particularly in the aircraft industry;
(5) cold-heading, as in the production of nails and bolts;
(6) coining and embossing.

6.40 Hot-working processes

A hot-working process is one which is carried out at a temperature well above the recrystallisation temperature of the metal or alloy. At such a temperature, recrystallisation will take place simultaneously with deformation, and so keep pace with the actual working process (fig. 6.7). For this reason, the metal will not work-harden, and can be quickly and

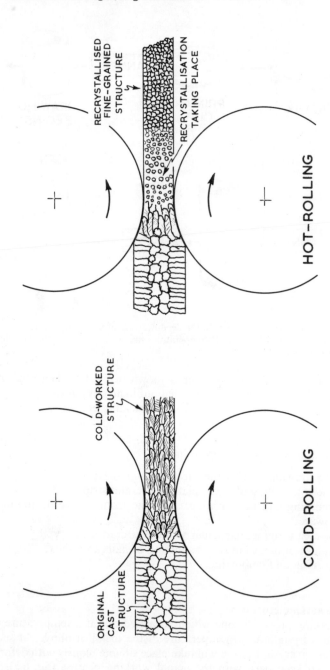

Fig. 6.7—During cold-rolling a metal becomes work-hardened, but during a hot-rolling process recrystallisation can take place.

continuously reduced to its required shape, with the minimum of expended energy. Not only is the metal naturally more malleable at a high temperature, but it remains soft, because it is recrystallising continuously during the working process.

6.41 Thus, hot-working leads to a big saving in both energy used and time required for a shaping process. It also results in the formation of a uniformly fine grain in the recrystallised material, replacing the original coarse cast structure. For this reason, the product is stronger, tougher, and more ductile than was the original cast material.

Condition	Tensile strength (N/mm²)	% elongation	Izod impact(J)
as cast	540	16	3
hot-rolled (in the direction of rolling)	765	24	100
hot-rolled (at right angles to the direction of rolling)	750	15	27

Table 6.2—The effects of hot-working on mechanical properties.

6.42 The main disadvantage of hot-working is that the surface condition is generally poor, due to oxidation and scaling at the high working temperature. Moreover, accuracy of dimensions is generally more difficult to attain, because the form tools need to be of simpler design for working at high temperatures. Consequently, hot-working processes are usually followed by some surface-cleaning process, such as acid-pickling, and at least one cold-working operation, which will improve the surface quality and accuracy of dimensions.

6.43 The principal hot-working processes are:

(1) hot-rolling, for the manufacture of plate, sheet, strip, and shaped sections such as rolled-steel joists;
(2) forging and drop-forging, for the production of relatively simple shapes, but with mechanical properties superior to those of castings;
(3) extrusion, for the production of many solid and hollow sections (tubes) in both ferrous and non-ferrous materials.

6.44 Many metal-working processes involve preliminary hot-working, followed by cold-forming or finishing. Mild-steel sheet destined for the production of a motor-car body is first hot-rolled down from the ingot to quite thin sheet. This is cooled, and then pickled in acid, after which it is lightly cold-rolled to give a dense smooth surface. Finally, it is cold-formed to shape by means of presswork. This final shaping also hardens and strengthens the material, making it sufficiently rigid for service. Similarly, a drop-forged connecting rod will be finally 'sized' by 'tapping' it in a cold die, in order to improve both dimensions and finish.

6.50 Grain flow and fibre

During any hot-working process, the metal is moulded in a plastic manner, with a result similar to that produced by a baker kneading his dough (though a machine now does this for him). Segregated impurities in the original casting are mixed in more uniformly, so that brittle films no longer coincide with grain boundaries. Obviously these impurities do not disappear completely, but form 'flow lines' or 'fibres' in the direction in which the material has been deformed. Since new crystals grow independently of these fibres, the latter weaken the structure to a much smaller extent than did the original intercrystalline films. Consequently, the material becomes stronger and tougher, particularly along the direction of the fibres. At right angles to the fibres, the material is weaker (table 6.2), since it tends to pull apart at the interface between the metal and each fibre.

6.51 Stages in the manufacture of a simple bolt are shown in fig. 6.8.

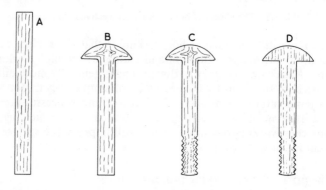

Fig. 6.8—The direction of 'fibre' in a cold-headed or forged bolt as compared with that in a bolt machined from a solid bar (D).

The bolt has been produced from steel rod in which a fibrous structure (A) is the result of the original hot-rolling process. The rod will be either hot- or cold-headed, and this operation will give rise to plastic flow in the metal, as indicated by the altered direction of the fibre in the head (B). The thread will then be rolled on, producing a further alteration in the direction of the fibre in that region (C). The mechanical properties of a bolt manufactured in this way will be superior to those of one machined from a solid bar (D). The latter would be much weaker, and it is highly likely that, in tension, the head may shear off, due to weakness along the flow lines—even supposing the thread had not already stripped for the same reason. For the mass production of bolts, the forging/thread-rolling method would be less costly in any case; so the choice of process here is very simple.

6.52 It is thus bad engineering practice to use any shaping process which exposes a cut fibrous structure such that shearing forces can act *along* exposed fibres, and so cause failure. Similarly, in a drop-forging it is essential that fibres are formed in those directions in which they will give rise to maximum strength and toughness.

6.60 The macro-examination of fibre direction
It is often necessary for the drop-forger to examine a specimen forging to ensure that grain flow is taking place in the desired direction. This will be indicated by fibre direction. Similarly, the engineer may wish to find out whether or not a component has been manufactured by forging—assuming that any such evidence as 'flash lines' and the like has been obliterated by subsequent light machining.

6.61 Whereas *micro*-examination involves the use of a microscope to view the prepared surface (10.50), *macro*-examination implies that no such equipment is necessary, other than, possibly, a hand magnifier. Hence the macro-examination of flow lines or fibre is a fairly simple matter. A section is first cut so that it will reveal fibre on a suitable face—in the case of the bolt mentioned above, a section cut symmetrically along its axis is used.

6.62 The section is then filed perfectly flat, or, better still, ground flat with a linishing machine. It is then rubbed on successively finer grades of emery paper, the 'grits' most generally used being '120', '180', '220', and '280'. Finer grades than these are not necessary, though the method of grinding requires a little care, to ensure that deep scratches are eliminated from the surface. This is achieved by grinding the surface such that, on passing from the linisher to the coarsest paper, the specimen is turned through 90°. The new set of grinding marks will then be at right angles to the previous ones, and in this way it is easy to see when the old grinding marks have been removed. The same procedure is adopted in passing from one paper to the next. Small specimens are ground by rubbing the specimen on paper supported on a piece of plate glass. Large specimens are more difficult to manipulate, and it is often more convenient to grind such specimens by rubbing a small area at a time, using a wooden block faced with emery paper.

6.63 When a reasonably smooth surface with no deep scratches has been produced, the specimen is wiped free of swarf and dust. It is then immersed in a 50% solution of hydrochloric acid, and gently heated. Considerable effervescence takes place as hydrogen gas is liberated, and the surface layers of steel are dissolved. Those areas containing the most impurity tend to dissolve more quickly, and this helps to reveal the fibrous structure. The specimen is examined from time to time, being lifted from the solution with laboratory tongs. This deep-etching process may take

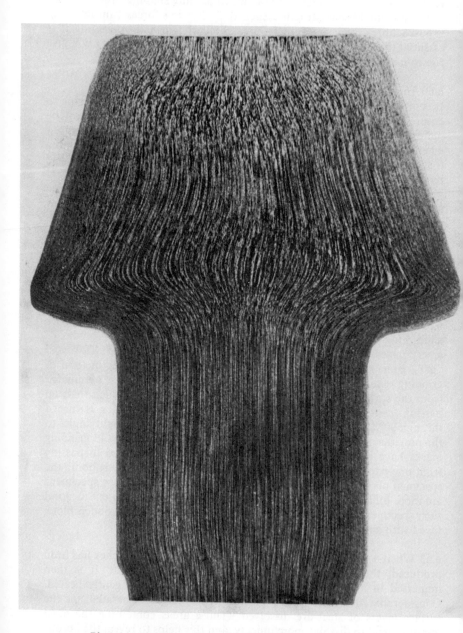

Plate 6.1—The fibrous structure of a hot-forged component (6.50).
The 'flow-lines' indicate the direction in which the metal moved during the forging
operation. Etched in boiling 50% hydrochloric acid for 15 minutes.

up to twenty minutes or so, depending upon the quality of the steel (Plate 6.1).

6.64 When the flow lines are suitably revealed, the specimen is washed in running water for a few seconds, and then dried as quickly as possible, by immersion in 'white' methylated spirit, followed by holding in a current of warm air from a hair-dryer.

6.65 If a section has been really deeply etched, it is possible to take an ink print of the surface. A blob of printer's ink is rolled on a flat glass plate, so that a thin but even film is formed on the surface of the rubber roller or squeegee used for this purpose. The roller is then carefully passed over the etched surface, after which a piece of paper is pressed on to the inked surface. Provided the paper does not slip in the process, a tolerable print of the flow lines will result.

Chapter Seven
The Mechanical Shaping of Metals

7.10 The craft of the smith was established quite early in Man's development. Some 6000 years ago, Tubal Cain was forging iron, no doubt using goat-skin bellows to raise the temperature of his hearth. We have every reason to believe that the smithy was established as a vital part of the community in much earlier days, and that, almost as soon as Man developed bronze, he began to exploit the malleability of those grades which were low in tin.

7.11 Modern shaping processes can be divided into hot-working operations and cold-working operations. The former tend to be used wherever possible, since less power is required, and working can be carried out more rapidly. Cold-working processes, on the other hand, are used in the final stages of shaping some materials; so that a high-quality surface finish can be obtained, or suitable strength and hardness developed in the material.

Hot-working processes
7.20 A hot-working process is one which is carried out at a temperature above that of recrystallisation for the material. Consequently, deformation and recrystallisation take place at the same time; so the material remains malleable during the working process. Intermediate annealing processes are therefore not required; so working takes place very rapidly.

7.21 *Forging*. The simplest and most ancient metal-working process is that of hand-forging, mentioned above. With the aid of simple tools called 'swages', the smith can produce relatively complex shapes, using either a hand- or a power-assisted hammer.

7.21.1 *Drop-forging*. If large numbers of identically shaped components are required, then it is convenient to mass-produce them by drop-forging. A shaped two-part die is used, one half being attached to the hammer, whilst the other half is carried by a massive anvil. For complicated shapes, a series of dies may be required.

The hammer, working between two vertical guides, is lifted either mechanically or by steam pressure, and is then allowed to fall, or is driven down (fig. 7.1) on to the metal to be forged. This consists of a hot bar of metal, held on the anvil by means of tongs. As the hammer comes into contact with the metal, it forges it between the two halves of the die.

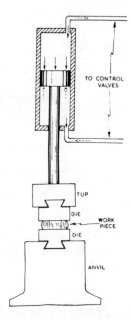

Fig. 7.1—A double-acting steam hammer.

7.21.2 *Hot-pressing.* This is a development of drop-forging which is generally used in the manufacture of simpler shapes. The drop hammer is replaced by a hydraulically driven ram; so that, instead of receiving a rapid succession of blows, the metal is gradually squeezed by the static pressure of the ram. The downwards thrust is sometimes as great as 500 MN.

The main advantage of hot-pressing over drop-forging is that mechanical deformation takes place more uniformly throughout the work-piece, and is not confined to the surface layers, as it is in drop-forging. This is important when forging large components like marine propellor-shafts, which may otherwise suffer from having a non-uniform internal structure.

7.22 *Hot-rolling.* Until the Renaissance in Europe, hand-forging was virtually the only method available for shaping metals. Rolling seems to have originated in France in about 1550; whilst in 1680 a sheet mill was in use in Staffordshire. In 1783, Henry Cort adapted the rolling mill for the production of wrought-iron bar. The introduction of mass-produced steel by Bessemer, in 1856, led to the development of bigger and faster rolling mills. The first reversing mill was developed at Crewe, in 1866, and this became standard equipment for the initial stages in 'breaking-down' large ingots to strip, sheet, rod, and sections.

A steel-rolling shop consists of a powerful 'two-high' reversing mill (fig. 7.2), to reduce the section of the incoming white-hot ingots, followed by trains of rolls which are either plain or grooved, according to the

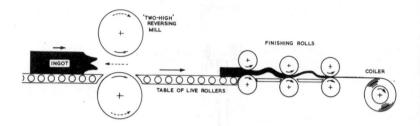

Fig. 7.2—The rolling of steel strip.

The ingot is first 'broken down' by the two-high reversing mill (the piped top is usually cropped after several passes through this mill). The work-piece is then delivered to the train of rolls, which roll it down to strip.

product being manufactured. Hot-rolling is also applied to most non-ferrous alloys in the initial stages of breaking-down, but finishing is more likely to be a cold-rolling operation.

7.23 *Extrusion*. The extrusion process is used for shaping a variety of both ferrous and non-ferrous alloys. The most important feature of the process is that, in a single operation from a cast billet, quite complex sections of reasonably accurate dimensions can be obtained. The billet is heated to the required temperature (350–500°C for aluminium alloys, 700–800°C for brasses, and 1100–1250°C for steels), and placed in the container of the extrusion press (fig. 7.3). The ram is then driven forward hydraulically, with sufficient force to extrude the metal through a hard alloy-steel die. The *solid* metal section issues from the die in a manner similar to the flow of tooth-paste from its tube.

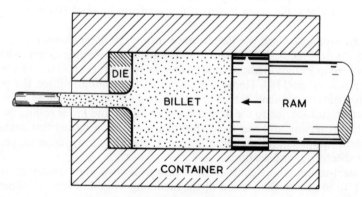

Fig. 7.3—The extrusion process.

A wide variety of sections can be extruded, including round rod, hexagonal brass rod (for parting off as nuts), brass curtain rail, small diameter rod (for drawing down to wire), stress-bearing sections in aluminium alloys (mainly for aircraft construction), and tubes in carbon and stainless steels, as well as in aluminium alloys and copper alloys.

Cold-working processes

7.30 The surface of a hot-worked component tends to be scaled, or at least heavily oxidised; so it needs to be sand-blasted or 'pickled' in an acid solution if its surface condition is to be acceptable. Even so, a much better surface quality is obtained if the component or material is cold-worked *after* being pickled. Consequently, some degree of cold-work is applied to most components as a final stage in manufacture. However, cold-working is also a means of obtaining the required mechanical properties in a material. By varying the amount of cold-work in the final operation, the degree of hardness and strength can be adjusted. Moreover, some operations can be carried out only on cold metals and alloys. Those processes which involve *drawing*—or pulling—the metal must generally be carried out on cold material, since ductility is usually less at high temperatures. This is because tensile strength is reduced; so the material tears apart very easily when heated. So, whilst malleability is increased by rise in temperature, ductility is generally reduced. Hence there are more cold-working operations than there are hot-working processes, because of the large number of different final shapes which are produced in metallic materials of varying ductility. Finally, cold-working allows much greater accuracy of dimensions to be obtained in the finished material.

7.31 *Cold-rolling.* Cold-rolling is used during the finishing stages in the production of both strip and section, and also in the manufacture of foils. The types of mill used in the manufacture of the latter are shown in fig. 7.4. To roll very thin material, small-diameter rolls are necessary; and, if the material is of great width, this means that the working rolls must be supported by backing rolls, otherwise they will bend to such an extent that reduction in thickness of very thin material becomes impossible. For rolling thicker material, ordinary two-high mills are generally used. The production of mirror-finished metal foil necessitates the use of rolls with a highly polished surface; only by working in perfectly clean surroundings with highly polished rolls can really high-grade foil be obtained.

7.32 *Drawing.* Drawing is exclusively a cold-working process, because it relies on the ductility of the material being drawn. Rod, wire, and hollow sections (tubes) are produced by drawing them through dies. In the manufacture of wire (fig. 7.5), the material is pulled through the die by winding it on to a rotating drum or 'block'; whilst in the production of tube, the bore is maintained by the use of a mandrel. Rods and tubes are drawn at a

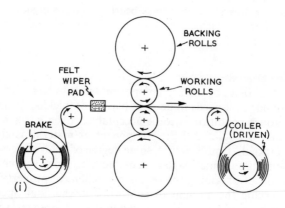

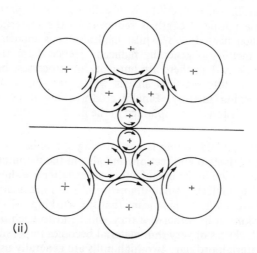

Fig. 7.4—Roll arrangements in mills used for the production of thin foil.
(i) A four-high mill.   (ii) A 'cluster' mill.

long draw-bench, on which a power-driven 'dog' pulls the material through the die (fig. 7.6). In each case, the material is lubricated with oil or soap before it enters the die aperture.

Drawing dies are made from high-carbon steel; from tungsten-molybdenum steels; from tungsten carbide; and, for very fine-gauge copper wire, from diamond.

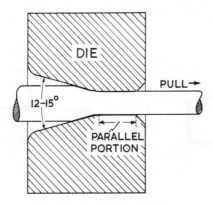

Fig. 7.5—A wire-drawing die.

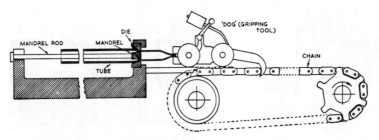

Fig. 7.6—A draw-bench for the drawing of tubes.
Rod could also be drawn at such a bench, the mandrel then being omitted.

7.33 *Cold-pressing and deep-drawing.* These processes are so closely allied to each other that it is often difficult to define each separately; however, a process is generally termed deep-drawing if some thinning of the walls of the component occurs under the application of tensile forces. Thus the operations range from making a pressing in a single-stage process to cupping followed by a number of drawing stages. In each case, the components are produced from sheet stock, and range from pressed mild-steel motor-car bodies to deep-drawn brass cartridge-cases, cupro-nickel bullet-envelopes, and aluminium milk-churns.

Only very ductile materials are suitable for deep-drawing. The best known of these are 70–30 brass (16.50), pure copper, cupro-nickel (16.80), pure aluminium and some of its alloys (17.50), and some of the high-nickel alloys. Mild steel of deep-drawing quality is now being produced in increasing quantities by the oxygen processes (11.22), and is used in a large number of motor-car parts.

Typical stages in a pressing and deep-drawing process are shown in fig. 7.7. Wall-thinning may or may not take place in such a process. If

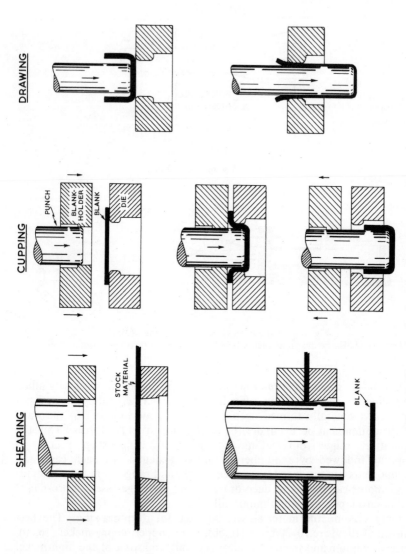

Fig. 7.7—Stages in the deep-drawing of a component.

wall-thinning is necessary, then a material of high ductility must be used. Although fig. 7.7 shows the processes of shearing and cupping being carried out on different machines, usually a combination tool is used; so that both processes take place on one machine.

Simple cold-pressing is widely used, and alloys which are not sufficiently ductile to allow deep-drawing are generally suitable for shaping by simple presswork. Much of the bodywork of a motor car is produced in this way, from mild steel.

7.34 *Spinning*. This is one of the oldest methods of shaping sheet metal, and is a relatively simple process, in which a circular blank of metal is attached to the spinning chuck of a lathe. As the blank rotates, it is forced into shape by means of hand-operated tools of blunt steel or hardwood, supported against a fulcrum pin (fig. 7.8). The purpose of the hand-tool is to press the

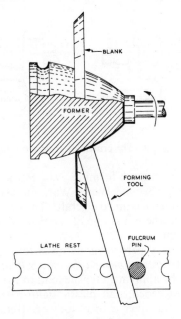

Fig. 7.8—Spinning.

metal blank into contact with a former of the desired shape. The former, which is also fixed to the rotating chuck, corresponds to the internal shape of the finished component, and may be made from a hardwood such as maple, or, in some cases, from metal. Formers may be solid; if the component is of re-entrant shape (as in fig. 7.8), then the former must be segmented, to enable it to be withdrawn from the finished product. Adequate lubrication is necessary during spinning. For small-scale work, beeswax or tallow are often used; whilst for larger work, soap is the usual choice.

Large reflectors, components used in chemical plant, stainless-steel dairy-utensils, aluminium teapots and hot-water bottles, ornaments in copper and brass, and other domestic hollow-ware are frequently made by spinning.

A mechanised process somewhat similar to spinning, but known as 'flow-turning', is used for the manufacture of such articles as stainless-steel or aluminium milk-churns. In this process, thick-gauge material is made to flow plastically, by pressure-rolling it in the same direction as the roller is travelling; so that a component is produced in which the wall thickness is much less than that of the original blank (fig. 7.9). Aluminium cooking-utensils in which the base is required to be thicker than the side walls are made in this way.

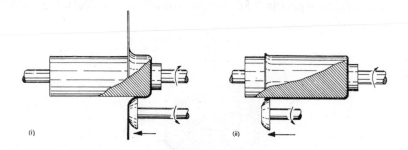

Fig. 7.9—The principles of flow-turning.

7.35 *Stretch-forming*. Stretch-forming was introduced in the aircraft industry just before the Second World War, and soon became important to the production of metal-skinned aircraft. The process is also used in the coach-building trade, where it is one of the principal methods of forming sheet metal and sections.

In any forming process, permanent deformation can only be achieved in the work-piece if it is *stressed beyond the elastic limit*. In stretch-forming, this is accomplished by applying a tensile load to the work-piece such that the elastic limit is exceeded, and plastic deformation takes place. The operation is carried out over a form-tool or stretch-block; so that the component assumes the required shape.

In the 'rising-table' machine (fig. 7.10), the work-piece is gripped between jaws, and the stretch-block is mounted on a rising table which is actuated by a hydraulic ram. Stretching forces of up to 4 MN are obtained with this type of machine. Long components, for example aircraft fuselage panels up to 7 m long and 2 m wide, are stretch-formed in a similar machine in which the stretch-block remains stationary, and the jaws move tangentially to the ends of the stretch-block.

Stretch-blocks are generally of wood or compressed resin-bonded ply-woods, though other materials, such as cast synthetic resins, zinc-base

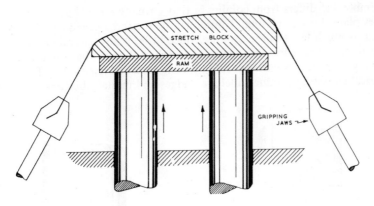

Fig. 7.10—The principles of stretch-forming.

alloys, or reinforced concrete, are also used. Lubrication of the stretch-block is, of course, necessary.

Although the process is applied mainly to the heat-treatable light alloys, stainless steel and titanium are also stretch-formed on a commercial scale.

7.36 *Coining and embossing.* Coining is a cold-forging process in which deformation takes place entirely by compression. It is confined mainly to the manufacture of coins, medals, keys, and small metal plaques. Frequently, pressures in excess of 1500 N/mm² are necessary to produce sharp impressions, and this limits the size of work which is possible.

The coining operation is carried out in a closed die (fig. 7.11). Since no provision is made for the extrusion of excess metal, the size of the blanks must be accurately controlled to prevent possible damage to the dies, due to the development of excessive pressures.

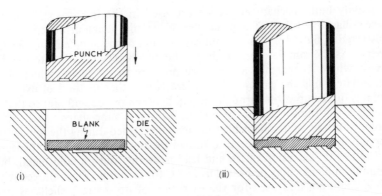

Fig. 7.11—Coining.

Embossing differs from coining in that virtually no change in thickness takes place during pressing. Consequently, the force necessary to emboss metal is much less than in coining, since little, if any, lateral flow occurs. The material used for embossing is generally thinner than that used for coining, and the process is effected by using male and female dies (fig. 7.12). Typical embossed products include badges and military buttons.

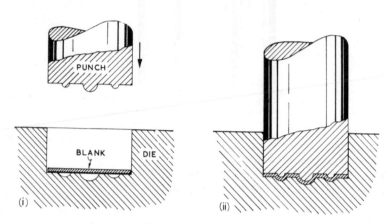

Fig. 7.12—Embossing.

7.37 *Impact-extrusion.* Extrusion as a hot-working process was described earlier in this chapter (7.23). A number of cold-working processes also fall under the general description of extrusion. Probably the best-known of these is the method by which disposable collapsible tubes are manufactured. Such tubes were produced in lead, for containing artists' colours, as long ago as 1841. A few years later, similar tubes were produced in tin, but it was not until 1920 that the impact-extrusion of aluminium was established on a commercial scale.

Heavily built mechanical presses are used in the impact-extrusion of these collapsible tubes. The principles of die and punch arrangement are illustrated in fig. 7.13. A small unheated blank of metal is fed into the die cavity. As the ram descends, it drives the punch very rapidly into the die cavity, where it transmits a very high pressure to the metal, which then immediately fills the cavity. Since there is no other method of exit, the metal is forced upwards through the gap between punch and die; so that it travels along the surface of the punch, forming a tube-shaped shell. The threaded nozzle of a collapsible tube may be formed during the impacting operation, but it is more usual to thread the nozzle in a separate process.

The impact-extrusion of tin and lead is carried out on cold metal, but aluminium blanks may be heated to 250°C for forming. Zinc, alloyed with 0·6% cadmium and used for the extrusion of dry-battery shells, is first heated to 160°C; so that the alloy becomes malleable.

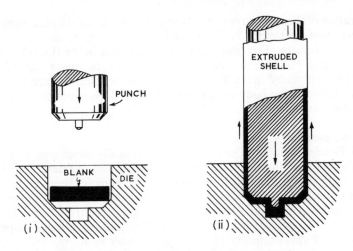

Fig. 7.13—The impact extrusion of a disposable tube.

Disposable tubes in lead, tin, and aluminium are used as containers for a wide range of substances, such as shaving cream, tooth-paste, medicines, shoe-polish, adhesives, and condensed milk. In addition to the manufacture of collapsible tubes, impact-extrusion is used for the production of many other articles, principally in aluminium. These include canisters and capsules for food, medical products, and photographic films; and shielding cans for radio components.

7.38 The Hooker process (fig. 7.14) closely resembles hot-extrusion, mentioned earlier, in so far as the flow of metal in relation to the die and punch is concerned. The Hooker process, however, is a cold-working

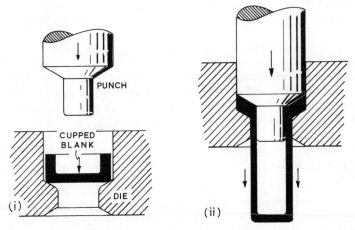

Fig. 7.14—Impact extrusion by the Hooker process.

operation of the impact type, and its products include small brass cartridge-cases, copper tubes for radiators and heat-exchangers, and other short tubular products. Flat 'slugs' are sometimes used in the Hooker process, but cupped blanks are usually considered to be more satisfactory. The blank is placed in the die; and, as the punch descends, metal is forced down between the body of the punch and the die, producing a tubular extrusion as shown.

Powder-metallurgy

7.40 Powder-metallurgy processes were originally developed to replace melting and casting for those metals—the so-called 'refractory' metals—which have very high melting-points. For example, tungsten melts at 3410°C, and this is beyond the softening temperature of all ordinary furnace-lining materials. Hence tungsten is produced from its ore as a fine powder. This powder is then 'compacted' in a die of suitable shape at a pressure of approximately 1500 N/mm². Under such high pressure, the particles of tungsten become joined together by 'cold-welding' at the points of contact between particles.

7.41 The compacts are then heated to a temperature above that of recrystallisation—about 1600°C in the case of tungsten. This treatment causes recrystallisation to occur, particularly in the highly deformed regions where cold-welding has taken place, and in this way the particles become joined, as grain-growth takes place across the original boundaries between particles. This heating process is known as 'sintering'.

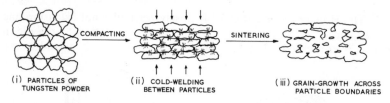

(i) PARTICLES OF TUNGSTEN POWDER   (ii) COLD-WELDING BETWEEN PARTICLES   (iii) GRAIN-GROWTH ACROSS PARTICLE BOUNDARIES

Fig. 7.15—Stages in a powder-metallurgy process.

At the end of this process, the slab of tungsten has a strong, continuous structure, though it will contain a large number of tiny cavities. It is then rolled, and drawn down to wire, which is used in electric lamp filaments. Most of the cavities are welded up by the working processes.

7.42 Although powder-metallurgy was originally used to deal with metals of very high melting-point, its use has been extended for other reasons, such as:

(1) to produce metals and alloys of *controlled* porosity, e.g. stainless-steel filters to deal with corrosive liquids, and also oil-less bronze bearings;

(2) to produce 'alloys' of metals which do not mix in the molten state, e.g. copper and iron for use as a cheap bearing material.

(3) for the production of small components such as the G-frame of a micrometer screw-gauge where the negligible amount of process scrap makes the method competitive.

7.43 One well known use of powder-metallurgy is in the manufacture of cemented carbides for use as tool materials. Here, tungsten powder is heated with carbon powder at about 1500°C, to form tungsten carbide. This is ground in a ball mill, to produce particles of very small size (about 20 μm), and the resultant tungsten carbide powder is mixed with cobalt powder; so that the particles of tungsten carbide become coated with powdered cobalt. The mixture is then compacted in hardened steel dies at pressures of about 300 N/mm², to cause cold-welding between the particles of cobalt. The compacts are then sintered at about 1500°C, to cause recrystallisation and grain growth in the cobalt; so that the result is a hard, continuous structure consisting of particles of very hard tungsten carbide in a matrix of hard, tough cobalt.

7.44 The principles involved in the manufacture of sintered-bronze bearings are slightly different. Here the main reason for using a powder-metallurgy process is to obtain a bearing with a controlled amount of porosity; so that it can be made to absorb lubricating oil. Powders of copper and tin are mixed in the correct proportions (about 90% copper, 10% tin), and are then compacted in a die of suitable shape (fig. 7.16). The compacts are then sintered at 800°C for a few minutes. This, of course, is *above* the melting-point of tin; so the process differs from true powder-

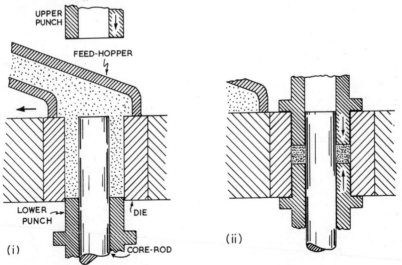

Fig. 7.16—The compacting process used in the manufacture of an 'oil-less' bronze bearing-bush.

metallurgy processes, in which no fusion occurs. As the tin melts, it percolates between the copper particles, and alloys with them to produce a continuous mass. The compact, however, still retains a large amount of its initial porosity; and, when it is quenched into lubricating oil, the latter is drawn into the pores of the bearing as it cools down. Sometimes the bearings are then placed under vacuum, whilst still in the oil bath. This causes any remaining air to be drawn out of the bearing, being replaced by oil when the pressure is allowed to return to that of the atmosphere.

7.45 The final structure resembles that of a metallic sponge which, when saturated with lubricating oil, produces a self-oiling bearing. In some cases, the amount of oil in the bearing lasts for the lifetime of the machine. Such bearings are used in the motor-car industry, but are also particularly useful in many domestic machines such as vacuum cleaners, refrigerators, electric clocks, and washing machines, in all of which long service with a minimum of attention is desirable.

Chapter Eight
Alloys

8.10 Most of our useful metals are soft and ductile when pure. Thus, pure copper and pure aluminium are admirable materials from which kettles, pots, pans and the like can be formed; whilst dead-mild steel—which industrially is the nearest feasible approach to pure iron—is widely used for the bodywork of motor cars, and countless other drawn or pressed components. Unfortunately, these pure metals, though ductile, are comparatively weak; and, whilst it is possible to increase their strengths by means of cold-work, it is usually necessary to obtain still greater strength and hardness by alloying.

An alloy is a mixture of two or more metals, made with the object of improving the properties of one of these metals, or, in some cases, producing new properties not possessed by either of the metals in the pure state. For example, pure copper has a very low electrical resistance, and is therefore used as a conductor of electricity; but, with 40% nickel, an alloy, 'Constantan', with a relatively high electrical resistance is produced. Again, pure iron is a ductile though rather weak material; yet the addition of less than 0·5% carbon will result in the exceedingly strong alloy we call steel. In this chapter we shall examine the internal structures of different types of alloy, and show to what extent the structures of these alloys influence their mechanical properties.

8.11 It is a general rule that, in order to produce a useful alloy, two metals must 'mix' with each other in the molten state. In some cases they do not; but, like oil and water, form two separate layers in their containing vessel. Clearly, such metals are unlikely to form a useful alloy, and we must therefore begin with the assumption that the two metals do mix; that is, they completely dissolve in each other in the liquid state, to form a single homogeneous solution. However, it is on the manner in which this liquid solution subsequently solidifies that we must focus our attention.

Eutectics
8.20 Sometimes on solidification, the two metals cease to remain dissolved in one another, but separate instead, each to form its own individual crystals. In a similar way, salt will dissolve in water; though when the solution reaches its freezing-point, individual crystals of pure ice and pure salt are formed. Another fact we notice is that the freezing-temperature of the salt solution is much lower than that of pure water (which is why the Local Authority scatters damp salt on our roads after a night of frost). This phenomenon is known as 'depression of freezing-point', and it is observed in the case of many metallic alloys. Thus, the addition of

increasing amounts of the metal cadmium to the metal bismuth will cause its freezing-point to be depressed proportionally; whilst, conversely, the addition of increasing amounts of bismuth to cadmium will have a similar effect on its melting-point, as shown in fig. 8.1.

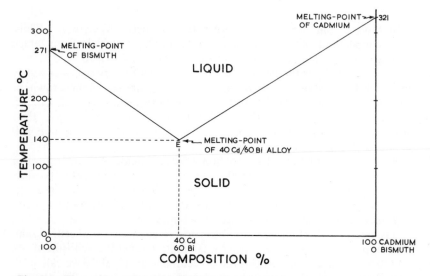

Fig. 8.1—The melting-points of both bismuth and cadmium are depressed by adding each to the other.
    A minimum freezing-point—or 'eutectic point'—is produced.

8.21 It will be noticed that the two lines meet at the point *E*, corresponding to an alloy containing 60% bismuth and 40% cadmium. This alloy melts and freezes at a temperature of 140°C, and is the lowest melting-point alloy which can be made by mixing the metals cadmium and bismuth. The point *E* is called the *eutectic* point*, and the composition 60% bismuth/ 40% cadmium is the *eutectic mixture*. If a molten alloy of this composition is allowed to cool, it will remain completely liquid until the temperature falls to 140°C, when it will solidify by forming alternating thin layers of pure cadmium and pure bismuth (fig. 8.2) until solidification is complete. The metallic layers in this eutectic structure are extremely thin, and a microscope with a magnification of at least 100 is generally necessary to be able to see the structure.

Since the structure is laminated, something like plywood, the mechanical properties of eutectics are often quite good. For example, when the material forming one type of layer is hard and strong, whilst the other is soft and ductile, the alloy will be characterised by strength and toughness, since the strong though somewhat brittle layers are cushioned between soft

---

* Pronounced 'you-tek-tick', this word is derived from the Greek *eutektos*, meaning 'capable of being melted easily'.

but tough layers. Thus the eutectic in aluminium-silicon alloys (17.61) consists of layers of hard, brittle silicon sandwiched between layers of soft, tough aluminium, and the tensile strength of these cast aluminium-silicon alloys is much higher than that of pure aluminium.

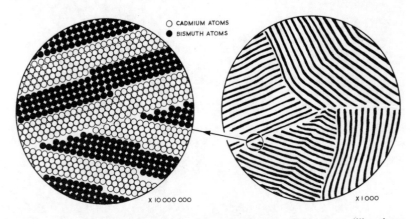

Fig. 8.2—If we had a microscope which gave a magnification of about ten million times, the arrangement of atoms of cadmium and bismuth would look something like that in the left-hand part of the diagram, except that the bands in the eutectic would each be many thousands of atoms in width.

Solid solutions

8.30 Sometimes two metals which are completely soluble in each other in the liquid state remain dissolved in each other during and after solidification, forming what metallurgists call a 'solid solution'. This is generally the case when the two metals concerned are similar in properties, and have atoms which are approximately equal in size. During solidification, the crystals which form are built from atoms of both metals. Inevitably, one of the metals will have a melting-point higher than that of the other, and it is reasonable to expect that this metal will tend to solidify at a faster rate than the one of lower melting-point. Consequently, the core of a resultant dendrite contains rather more of the metal with the higher melting-point; whilst the outer fringes of the crystal will contain correspondingly more of the metal with the lower melting-point (fig. 8.3). This effect, known as 'coring', is prevalent in all solid solutions in the cast condition.

8.31 Many students appear to find the basic idea of solid solution one which is difficult to understand; so an analogy, in which bricks replace atoms as the building units, will be drawn here.

Suppose that, in building a high wall, a team of bricklayers is given a mixture of red and blue bricks with which to work. Further, let us suppose that, as building proceeds, each successive load of bricks which arrives at

Plate 8.1—The dendritic structure of a cast alloy.
This is a photomicrograph of cast 70–30 brass at a magnification of ×50. The dendrites would not be visible were it not for the coring of the solid solution (8.30).

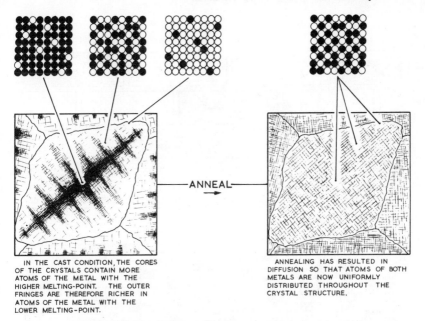

IN THE CAST CONDITION, THE CORES OF THE CRYSTALS CONTAIN MORE ATOMS OF THE METAL WITH THE HIGHER MELTING-POINT. THE OUTER FRINGES ARE THEREFORE RICHER IN ATOMS OF THE METAL WITH THE LOWER MELTING-POINT.

ANNEALING HAS RESULTED IN DIFFUSION SO THAT ATOMS OF BOTH METALS ARE NOW UNIFORMLY DISTRIBUTED THROUGHOUT THE CRYSTAL STRUCTURE.

● ATOMS OF THE METAL WITH THE HIGHER MELTING-POINT
○ ATOMS OF THE METAL WITH THE LOWER MELTING-POINT

Fig. 8.3—The variations in composition in a cored solid solution. The coring can be dispersed by annealing.

the site contains a slightly higher proportion of red bricks than the previous load. We will also assume that the 'brickies', being paid on a piece-work basis, lay whatever brick comes to hand first. Thus there will be no pattern in the laying of individual bricks; though, as the wall rises, there will be more red bricks and less blue ones in successive courses. By standing close to the wall, one would observe a small section, possibly like that shown in fig. 8.4 (i), and this would give no indication of the overall distribution of red and blue bricks in the wall. On standing further away from the wall, the lack of any pattern in the laying of individual bricks would still be obvious, though the general relationship between the numbers of red and blue bricks at the top and bottom of the wall would become apparent (fig. 8.4(ii)). Provided that the red and blue bricks were of roughly equal size and strength, the wall would be perfectly sound, though it might look rather odd. Let us assume that we now view the wall from a distance of about half a kilometre. Individual bricks will now no longer be visible, though a gradual change in colour from blue at the bottom, through various shades of purple, to red at the top will indicate the relative numbers of each type of brick at various levels in the wall (fig. 8.4 (iii)).

8.32 The distribution of the two different types of atom in many solid

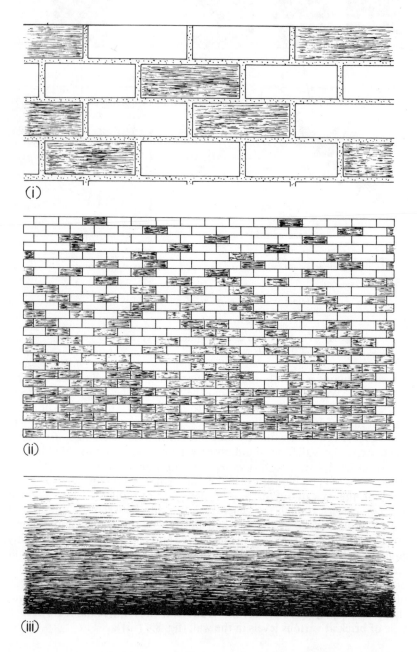

(i)

(ii)

(iii)

Fig. 8.4—The 'brick-wall analogy' of a cored solid solution.

solutions follows a pattern similar to that of the bricks in the rather curious wall just described; that is, the atoms in general conform to some overall pattern—as do bricks in the wall—but there is usually no rule governing the order in which single atoms of different types will arrange themselves within that pattern. The brick-wall analogy also illustrates the folly of viewing a microstructure at a high magnification, without first examining it with a low-power lens in the microscope. When using a high magnification, one will have only a restricted field of view of a very small part of the structure, so that no overall pattern is apparent; whereas the use of a low-power lens may reveal the complete dendritic structure, and show beyond doubt the nature of the material.

8.33 In the above discussion, it was assumed that the type of solid solution formed was one in which the atoms of the two metals concerned were of roughly equal size; and this type of structure (fig. 8.5 (i)) is termed a *substitutional* solid solution, since atoms of one metal have, so to speak, been substituted for atoms of the other. Many pairs of metals, including copper/nickel, silver/gold, chromium/iron, and many others, form solid solutions of this type in all proportions. A still greater number of pairs of metals will dissolve in each other in this way but in limited proportions. Notable examples include copper/tin, copper/zinc, copper/aluminium, aluminium/magnesium, and a host of other useful alloys.

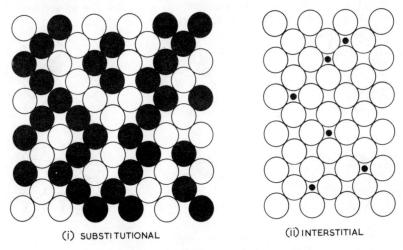

(i) SUBSTITUTIONAL   (ii) INTERSTITIAL

Fig. 8.5—The two main types of solid solution:
(i) a substitutional solid solution,
(ii) an interstitial solid solution.

8.34 There is, however, another type of solid solution, which is formed when the atoms of one element are so much smaller than those of the other that they are able to fit into the *interstices* (or spaces) between the larger atoms. Accordingly, this is known as an *interstitial solid solution* (fig. 8.5 (ii)). Carbon dissolves in face-centred cubic iron in this way. Since the relatively small carbon atoms fit into the spaces between the much larger iron atoms, this explains why a piece of *solid* steel can be carburised by being heated in the presence of carbon at a temperature high enough to make the steel face-centred cubic in structure. The carbon atoms 'infiltrate', so to speak, through the face-centred cubic ranks of iron atoms.

8.35 When a solid solution of either type is heated to a temperature which is high enough, some diffusion takes place, and the coring gradually disappears as the structure becomes more uniform in composition throughout. This is achieved by the movement of atoms from one part of a crystal to another. It is easy to see that this can happen in an interstitial solid solution, but in a substitutional solid solution it can occur only if we assume that gaps where atoms are missing exist in the crystals, thus enabling those atoms present to change places with each other. Examination of a piece of cast metal under the microscope will generally reveal the presence of a few minute cavities, even in good quality material. Even the smallest visible cavity represents a gap from which many thousands of atoms are missing; so it is reasonable to assume that there are countless small cavities representing a few missing atoms in each case, and which are far too small to be seen by the most powerful electron microscope. The presence of such cavities (or 'vacant sites' as they are called) will allow the movement of individual atoms to take place in substitutional solid solutions.

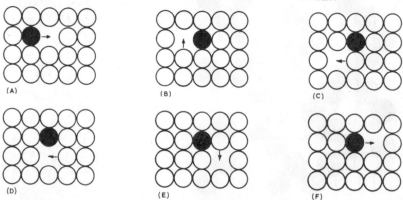

Fig. 8.6—This is the way in which metallurgists believe diffusion takes place in a substitutional solid solution. A series of five 'moves' is necessary in order that the 'black' atom (and its accompanying 'vacant site') can go forward by one space.

Metallurgists believe that this process of diffusion takes place in a manner similar to that shown in fig. 8.6. It is reasonable to suppose that, since the solute* atom will be either larger or smaller than the solvent atoms, it

* A *solution* consists of a *solute* dissolved in a *solvent*.

will possibly arrive alongside a vacant site, because in this way stress in the structure will be kept at a minimum. If the reader is good at chess— or even at draughts—he will be able to follow the 'moves' (as suggested in fig. 8.6) which have enabled the 'black' solute atom to advance to the right by one space. The series of 'moves' will, of course, be repeated every time the solute atom and its accompanying vacant site move forward by one space.

8.36 Solid solutions are possibly the most useful of metallurgical structures, since they generally give a combination of strength, toughness, and ductility. The increase in strength is due to the fact that, when two different metals crystallise as a solid solution, there is generally some distortion of the structure, due to the different sizes of the atoms involved. Slip (6.10) then becomes more difficult, and a greater force must be applied to produce it—which is another way of saying that the yield stress has been increased. Since slip is nevertheless still ultimately able to take place, much of the ductility of the original pure metal is retained. Most of our useful metallic alloys are basically solid solution in structure.

Intermetallic compounds
8.40 When heated, many metals combine with oxygen to form compounds which we call oxides; whilst some metals are attacked by sulphur gases in furnace atmospheres, to form sulphides. This is a general pattern in which metals (said to be electropositive elements) combine with non-metals (said to be electronegative substances) to form compounds which usually bear no physical resemblance to the elements from which they are formed. Thus, the very reactive metal sodium (which is silvery-white in appearance) combines with the very reactive gas chlorine (which is greenish in colour) to form sodium chloride—alias table-salt.

8.41 Sometimes two metals, when melted together, will combine to form a definite chemical compound called an 'intermetallic compound'. This often happens when the two metals are very unlike in their physical and chemical properties, and when one metal is strongly electropositive and the other weakly electropositive.

When a solid solution is formed, it usually bears at least some resemblance to the parent metals, as far as colour and other physical properties are concerned. Thus, the colour of a low-tin bronze (16.60) is a blend of the colours of its parent metals, copper and tin, as one might expect; but if the amount of tin is increased, so that the alloy contains 66% copper and 34% tin, a hard and extremely brittle substance is produced, bearing no resemblance whatever to either copper or tin. What is more, this intermetallic compound—for such it is—is of a pale blue colour. This is due mainly to the fact that an intermetallic compound generally crystallises in a different pattern to that of either of the parent metals. Moreover, since an intermetallic compound is always of fixed composition, in common with all chemical compounds, there is never any coring in crystals of such a substance in the cast state.

8.42 Because of the excessive brittleness of most of these intermetallic compounds, they are used to only a limited extent as constituents of engineering alloys, and then only in the form of small, isolated particles in the microstructure. Since many intermetallic compounds are very hard, they also have very low coefficients of friction. Consequently, one of their principal uses is as a constituent of bearing metals (19.60). If present in an alloy in large amounts, an intermetallic compound will often form brittle intercrystalline networks. The strength and toughness of such an alloy would be negligible, and it would be of no use to the engineer.

8.50 The entities described in this chapter are virtually the basic units from which metallic alloys are composed. It may be helpful, therefore, to summarise their properties.

(1) *Solid solutions* are formed when one metal is very similar to another, both physically and chemically, and is able to replace it, atom for atom, in its crystal structure. Alternatively, if the atoms of the second element are very small, they may be able to fit into the spaces between the larger atoms of the other metal.

Solid solutions are stronger than pure metals, because the presence of atoms of the second metal causes some distortion of the crystal structure, thus making slip more difficult. At the same time, solid solutions retain much of the toughness and ductility of the original pure metal.

(2) *Intermetallic compounds* are formed by chemical combination, and the resultant substance generally bears little resemblance to its parent metals. Most intermetallic compounds are hard and brittle, and are of limited use only in engineering alloys.

(3) *Eutectics* are formed when two metals, soluble in each other in the liquid state, become insoluble in each other in the solid state. Then alternate layers or bands of each metal form, until the alloy is completely solid. This occurs at a fixed temperature, which is below the melting-points of either of the two pure metals.

When two metals are only partially soluble in each other in the solid state, a eutectic may form consisting of alternate layers of two solid solutions. In some cases, a eutectic may consist of alternate layers of a solid solution and an intermetallic compound.

Chapter Nine
Equilibrium Diagrams

9.10 Among engineering students generally, the subject of equilibrium diagrams seems to have the reputation of being one which is best avoided if possible. Nevertheless, the topic need cause the reader no undue alarm, since for most purposes we can regard the equilibrium diagram as being no more than a graphical method of illustrating the relationship between the composition, temperature, and structure, or state, of any alloy in a series.

Much useful information can be obtained from these diagrams, if a simple understanding of their meaning has been acquired. For example, an elementary knowledge of the appropriate equilibrum diagram enables us to decide upon a suitable heat-treatment process to produce the required properties in a carbon steel. Similarly, a glance at the equilibrium diagram of a non-ferrous alloy system will often give us a pretty good indication of the structure—and hence the mechanical properties—a particular composition is likely to have. In attempting to assess the properties of an unfamiliar alloy, the modern metallurgist invariably begins by consulting the thermal equilibrium diagram for the series. There is no reason why the engineering technician should not be in a position to do precisely the same.

9.11 How are these equilibrium diagrams devised? Purely by much laborious experimental work, accompanied by experience of the behaviour of alloys; and in some cases, one suspects, by a certain amount of inspired guesswork.

The reader may already know that the composition of an ordinary tin/lead solder is chosen to suit the use to which it is to be put. Plumber's solder contains roughly two parts by weight of lead to one of tin. On cooling, it begins to solidify at about 265°C, but is not completely solid until its temperature has fallen to 183°C (fig. 9.2 (i)). It thus passes through a mushy, or pasty, part solid/part liquid range of some 80°C, enabling the plumber to 'wipe' a joint in a fractured domestic lead pipe. Even in these days of copper (or polythene) plumbing, sufficient lead water-piping still remains to keep a small army of plumbers* busy after a severe winter frost.

In the case of tinman's solder, used to join pieces of suitable metal, rather different properties are required. The solder must of course 'wet' (alloy with) the surfaces to be joined; but it will be an advantage if its

---

* A 'plumber' was originally one who worked in lead—the word is derived from the Latin *plumbum*, lead.

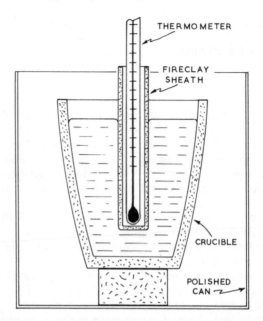

Fig. 9.1—Simple apparatus for determining the freezing-range of a low-temperature alloy.
The polished can prevents the alloy from cooling too quickly, and also protects the crucible from draughts.

melting-point is low, and, more important still, if it freezes quickly over a small range of temperature, so that the joint is less likely to be broken by rough handling in its mushy stage. A solder with these properties contains 62% tin and 38% lead. It freezes, as does a pure metal, at a single temperature—183°C in this case (fig. 9.2 (ii)). Since the cost of tin is more than ten times that of lead, tinman's solder often contains less than the ideal 62%. It will then freeze over a range of temperature which will vary with

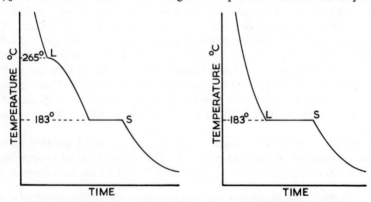

Fig. 9.2—Cooling curves for plumber's solder and for tinman's solder.

its composition. Thus, 'coarse' tinman's solder contains 50% tin and 50% lead. It begins to solidify at 220°C, and is completely solid at 183°C.

9.12 The freezing-range of any tin-lead solder can be determined by melting a small amount of it in a clay crucible, and then taking temperature readings of the cooling alloy every 15 seconds (fig. 9.1). A thermocouple is probably the best temperature-measuring instrument to use for this; though a '360°C' thermometer will suffice, provided it is protected by a fireclay sheath. Failure to use the latter will probably lead to the fracture of the thermometer as the solder freezes on to it, contracting in the process.

A temperature/time cooling curve can now be plotted in order to determine accurately the temperature at which freezing of the alloy begins (L) and finishes (S) (fig. 9.2).

This procedure is repeated for a number of tin-lead alloys of different compositions. Representative values of L and S for some tin-lead alloys are shown in the following table.

Composition		Temperature at which solidification begins (L) °C	Temperature at which solidification ends (S) °C
Lead %	Tin %		
67	33	265	183
50	50	220	183
38	62	183	183
20	80	200	183

The information obtained from the above table can now be plotted on a single diagram, as shown in fig. 9.3, in order to relate freezing-range to composition of alloy.

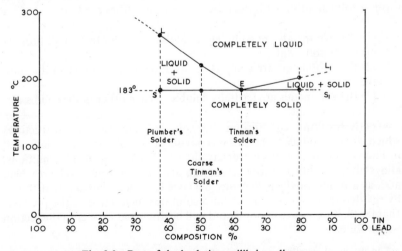

Fig. 9.3—Part of the lead–tin equilibrium diagram.
The limited information obtained from the cooling curves mentioned enables us to construct only so much of the diagram.

The line LEL_1 (called the *liquidus*) joins all points (L) at which solidification of the various alloys begins; whilst the line SES_1 (called the *solidus*) joins all points (S) at which solidification of the alloys has finished. What we have plotted is only a *part* of the lead-tin thermal equilibrium diagram. The complete diagram contains other lines—or 'phase*-boundaries', as they are called. To determine these lines, other, more complex methods have to be used; but we are not concerned here with advanced metallurgical laboratory practice.

9.13 Even this small portion of the lead-tin equilibrium diagram provides us with some useful information. We can, for example, use it to determine the freezing-range of a lead-tin alloy of given composition. Thus (reading from the diagram), a solder containing 60% lead and 40% tin will solidify between 250°C and 183°C. Similarly, given the composition of an alloy and its temperature at any instant, we can determine the state in which it exists. For example, an alloy containing, say, 55% lead and 45% tin, and at a temperature of 200°C, will be in a pasty, part solid/part liquid state; whilst the same alloy at 250°C will be completely molten. Conversely, when cooled below 183°C, it will be completely solid.

Types of equilibrium diagram
9.20 There are a number of different types of thermal equilibrium diagrams, but we need deal with only three of them, namely those in which the characteristics of the diagram are governed by the extent to which one metal forms a solid solution with the other. The possibilities are that:

(1) the two metals are completely soluble in each other in all proportions in the solid state,
(2) the two metals are completely insoluble in each other in the solid state,
(3) the two metals are partially soluble in each other in the solid state.

Strictly speaking, equilibrium diagrams indicate only microstructures which will be produced when alloys cool under equilibrium conditions, and in most cases that means extremely slowly. Under industrial conditions, alloys often solidify and cool far too rapidly for equilibrium to be reached, and, as a result, the final structure deviates considerably from that shown by the diagram. The coring of solid solutions mentioned in the previous chapter (8.30) is a case in point, and will be discussed further in the section which follows.

* In structural metallurgy, a 'phase' refers to any *single* substance which is present in an alloy system. Thus a phase may be either a pure metal, a solid solution, an intermetallic compound; or, in fact, a uniform liquid containing one or more metals.

An alloy system in which the two metals are soluble in each other in all proportions in both liquid and solid states

9.30 An example of this type of system is afforded by the nickel-copper alloy series. Atoms of nickel and copper are of approximately the same size, and, since both metals crystallise in similar face-centred cubic patterns (2.13), it is not surprising that they form mixed crystals of a substitutional solid-solution type when a liquid solution of the two metals solidifies. The resulting equilibrium diagram (fig. 9.4) will have been derived from a series of cooling curves, as described earlier in this chapter, except that a pyrometer capable of withstanding high temperatures would be required for taking the temperature measurements.

9.31 This equilibrium diagram consists of only two lines:

(1) the upper, or liquidus, above which any point represents in composition and temperature an alloy in the completely molten state; and

(2) the lower, or solidus, below which any point represents in composition and temperature an alloy in the completely solid state.

Any point between the two lines will represent in composition and temperature an alloy in the pasty or part solid/part liquid state. From the diagram, we can read not only the compositions of the solid part and liquid part respectively, but also determine the relative proportions of the solid and liquid material.

9.32 Let us consider what happens when an alloy (X), containing 60% nickel and 40% copper, cools and solidifies *extremely slowly*; so that its structure is able to reach equilibrium at every stage of the process.

Solidification will begin when the temperature falls to T (the vertical line representing the composition X and the horizontal line representing the temperature T intersect on the liquidus line). Now, it is a feature of equilibrium diagrams that, when a horizontal line representing some temperature cuts two adjacent phase-boundaries in this way, the compositions indicated by those two intersections can exist in equilibrium together. In this case, liquid solution of composition X can exist in equilibrium with solid solution of composition Y at the temperature T. Consequently, when solidification begins, crystal nuclei of composition Y begin to form.

9.33 Since the solid Y contains approximately 92% nickel/8% copper (as read from the diagram), it follows that the liquid which remains will be less rich in nickel (but correspondingly richer in copper) than the original 60% nickel/40% copper composition. In fact, as the temperature falls slowly, solidification continues; and *the composition of the liquid changes along the liquidus line; whilst the composition of the solid changes—by means of diffusion (8.35)—along the solidus line.* Thus, by the time the temperature has fallen to T_1, the liquid solution has changed in composition to X_1; whilst the solid solution has changed in composition to Y_1.

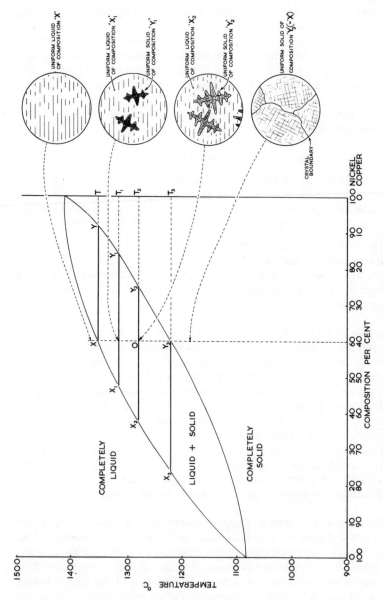

Fig. 9.4—The nickel-copper thermal equilibrium diagram.
The solidification of an alloy under conditions of equilibrium (slow cooling) is illustrated.

At a lower temperature—say T_2—solidification will have progressed still further; and the composition of the liquid will have changed to X_2; whilst the composition of the corresponding solid will have changed to Y_2.

9.34 Clearly, since the alloy is gradually solidifying so that the *proportions* of solid and liquid are continually changing, the *compositions* of solid and liquid must also change, because the overall composition of the alloy as a whole remains at 60% nickel/40% copper throughout the process. The relative weights of solid and liquid—as well as their compositions—can be obtained from the diagram, assuming that the alloy is cooling slowly enough for equilibrium to be attained by means of diffusion. Thus, at temperature T_2,

weight of liquid (composition X_2) . OX_2
$$= \text{weight of solid (composition } Y_2) . OY_2$$

This is commonly referred to as the *lever rule*. Engineers will appreciate that this is an apt title, since, in this particular case, it is as though moments had been taken about the point O. We will now substitute actual values (read from the equilibrium diagram, fig. 9.4) in the above expression. Then,

weight of liquid (38% nickel/62% copper) . (60 − 38)
$$= \text{weight of solid (74% nickel/26% copper) . (74 − 60)}$$

or

$$\frac{\text{weight of liquid (38\% nickel/62\% copper)}}{\text{weight of solid (74\% nickel/26\% copper)}} = \frac{(74 - 60)}{(60 - 38)} = \frac{14}{22}$$

Thus, assuming that the alloy is cooling slowly, and is therefore in equilibrium, we can obtain the above information about it at the temperature T_2 (1280°C).

The solidification process will finish as the temperature falls to T_3. Here the last trace of liquid (X_3) will have been absorbed into the solid solution, which, due to diffusion, will now be of uniform composition Y_3.

9.35 Composition Y_3 is of course the same as X—the composition of the original liquid. Obviously, it cannot be otherwise if the solid Y_3 has become uniform throughout due to diffusion. 'Why go through all this complicated procedure to demonstrate an obvious point?', the reader may ask. Unfortunately—whether in engineering or in other branches of applied science—it is not often possible to make a straightforward application of a simple scientific principle. Influences of other variable factors usually have to be taken into account, such as the effects of friction in a machine, or of the pressure of wind in a civil engineering project.

9.36 In the above application of the solidification of the 60% nickel/40% copper alloy, we have assumed that diffusion has taken place completely, resulting in the formation of a *uniform* solid solution. Under industrial conditions of relatively rapid cooling this is rarely possible: there just isn't time for the atoms to 'jiggle' around as described in the previous chapter (8.35). Consequently, the composition of the solid solution always lags behind that indicated by the equilibrium diagram for some particular temperature, and this leads to some residual coring in the final solid. If the rate of solidification has been very rapid, the cores of the crystals may be of a composition almost as rich in nickel as Y; whilst the outer fringes of the crystals may be of a composition somewhere in the region of X_3. Slower rates of solidification will lead to progressively lesser degrees of coring, as, under these circumstances, the effects of diffusion make themselves felt. Alternatively, if this 60% nickel/40% copper alloy were annealed for some hours at a temperature just below T_3, that is, just below the solidus temperature, any coring would be dissipated by diffusion.

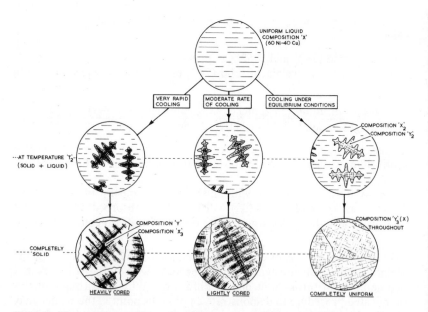

Fig. 9.5—Illustrating the effects of cooling-rate on the extent of coring in the 60% nickel/40% copper alloy dealt with above.

9.37 In this section we have been dealing with different modes of solidification of a 60% nickel/40% copper alloy. Since nickel and copper are soluble in each other in all proportions in the solid state, any other alloy composition of these two metals will solidify in a similar manner.

An alloy system in which the two metals are soluble in each other in all proportions in the liquid state, but completely insoluble in the solid state

9.40 In this case, the two metals form a single homogeneous liquid when they are melted together, but on solidification they separate again and form individual crystals of the two *pure* metals. Cadmium and bismuth form alloys of this type. Both metals have low melting-points, but, whilst cadmium is a malleable metal used to some extent for electroplating, bismuth is so brittle as to be useless for engineering purposes. It should be noted that the name 'bismuth' is often used to describe a compound of the actual metal which is sometimes used in medicine.

9.41 Again, the equilibrium diagram consists of only two boundaries: the liquidus *BEC*, and the solidus *AED*—or, more properly, *BAEDC*. As in the previous case, any point above *BEC* represents in composition and temperature an alloy in the completely molten state; whilst any point below *AED* represents an alloy in the completely solid state. Between *BEC* and *AED*, any point will represent in composition and temperature an alloy in the part liquid/part solid state.

9.42 Let us consider a molten alloy of composition X, containing 80% cadmium and 20% bismuth. This will begin to solidify when the temperature falls to T (fig. 9.6). In this case, the appropriate 'temperature horizontal' through T cuts that part of the solidus *BA* which represents a composition of 100% cadmium. Consequently, nuclei of pure cadmium begin to crystallise. As a result, the remaining molten alloy is left less rich in cadmium and correspondingly richer in bismuth; so, as the temperature falls, and cadmium continues to solidify, the liquid composition follows the liquidus line from X to X_1. This process continues, and, by the time the temperature has fallen to T_2, the remaining liquid will be of composition X_2.

The crystallisation of pure cadmium continues in this manner until the temperature has fallen to 140°C (the final solidus temperature), when the remaining liquid will be of composition E (40% cadmium/60% bismuth). Applying the lever rule at this stage:

weight of pure cadmium . AO = weight of liquid (composition E) . OE

or
$$\frac{\text{weight of pure cadmium}}{\text{weight of liquid (composition } E)} = \frac{EO}{AO}$$

$$= \frac{(60 - 20)}{(20 - 0)}$$

$$= \frac{40}{20}$$

$$= 2$$

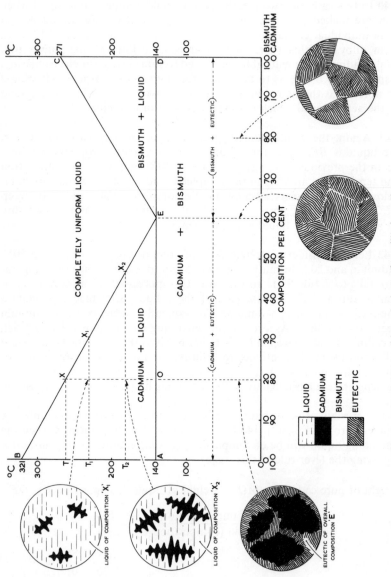

Fig. 9.6—The cadmium–bismuth thermal equilibrium diagram.

Thus, there will be twice as much solid cadmium by weight as there is liquid at this stage.

9.43 The two metals are now roughly in a state of equilibrium in the remaining liquid, which is represented in composition and temperature by the point E (the eutectic point). Until this instant, the liquid has been adjusting its composition by rejecting what are usually called 'primary' crystals of cadmium. However, due to the momentum of solidification, a little too much cadmium solidifies, and this causes the composition of the liquid to swing back across point E, by depositing for the first time a thin layer of bismuth. Since this now upsets equilibrium in the other direction, a layer of cadmium is deposited, and so the liquid composition continues to swing to-and-fro across the eutectic point, by depositing alternate layers of each metal until the liquid is all used up. This see-saw type of solidification results in the laminated structure of eutectics (8.21), and takes place whilst the temperature remains constant—in this case at the eutectic temperature of 140°C.

9.44 It was shown above that, just before solidification of the eutectic began,

weight of pure cadmium $= 2$. (weight of liquid of composition E)

Since the liquid of composition E has changed to eutectic, it follows that the final structure will contain two parts by weight of primary cadmium to one part by weight of eutectic. By the same reasoning, an alloy containing 70% cadmium/30% bismuth will contain equal parts of primary cadmium and eutectic; whilst an alloy containing 40% cadmium/60% bismuth will consist entirely of eutectic.

Alloys containing more than 60% bismuth will begin to solidify by depositing primary crystals of bismuth, and the general procedure will be similar to that outlined above for the cadmium-rich alloy. Whatever the *overall* composition of the alloy, the eutectic it contains will always be of the same composition; that is, 40% cadmium/60% bismuth. If the overall composition contains more than 40% cadmium, then some primary cadmium must deposit first; whilst if the overall composition has less than 40% cadmium, then some primary bismuth will deposit first.

An alloy system in which the two metals are soluble in each other in all proportions in the liquid state, but only partially soluble in each other in the solid state
9.50 This is a state of affairs which is intermediate between the two previous cases, since it represents a compromise between the extremes of complete solid solubility on the one hand and complete insolubility in the solid state on the other. As might be expected, very many alloy systems fall between these two extremes, and are therefore represented by this case. In the early paragraphs of this chapter, part of the lead-tin thermal

equilibrium was used to give a general idea of a method by which these thermal equilibrium diagrams can be produced. We shall now explore the lead-tin system more fully, by reference to the complete equilibrium diagram (fig. 9.7).

This diagram indicates that at 183°C lead will dissolve a maximum of 19·5% tin in the solid state, giving a solid solution designated α*; whilst, at the same temperature, tin will dissolve a maximum of 2·6% lead, forming a solid solution β. Any alloy whose composition falls between B and F will show a structure consisting of primary crystals of either α or β, and also some eutectic of α and β. The overall composition of the eutectic part of the structure will be given by E. In fact, an alloy containing precisely 62% tin and 38% lead will have a structure which is entirely eutectic, consisting of alternate layers of α and β.

9.51 Let us consider what happens when an alloy containing 70% lead and 30% tin solidifies, and cools *slowly* to room temperature. Solidification will commence at X (at about 270°C), when nuclei of α (composition Y) begin to form. By the time the temperature has fallen to, say, 220°C, the α will have changed in composition to Y_1, due to diffusion (8.35); whilst the remaining liquid will be of composition X_1. By applying the lever rule we have that, at 220°C,

weight of solid α (composition Y_1) . Y_1P
$$= \text{weight of liquid (composition } X_1) . PX_1$$

or $\quad \dfrac{\text{weight of solid α (composition } Y_1)}{\text{weight of liquid (composition } X_1)} = \dfrac{PX_1}{Y_1P}$

Similarly, when the temperature has fallen to 183°C, we have α (now of composition B) and some remaining liquid (of composition E) in a ratio given by:

$$\frac{\text{weight of solid α (composition } B)}{\text{weight of liquid (composition } E)} = \frac{QE}{BQ}$$

At a temperature just below 183°C, the remaining liquid solidifies as a eutectic, by depositing alternate layers of α (composition B) and β (composition F), the *overall* composition of this eutectic being given by E. Thus the structure, represented by a point just below Q, will consist of primary crystals of α of uniform composition B, surrounded by a eutectic mixture of α (composition B) and β (composition F).

9.52 In this diagram, we have two phase boundaries of a type not previously encountered in our studies, namely the *solvus* lines BC and FG. These boundary lines separate phase fields in which only solid phases

---

* Metallurgists use letters of the Greek alphabet to indicate different solid phases occurring in an alloy system.

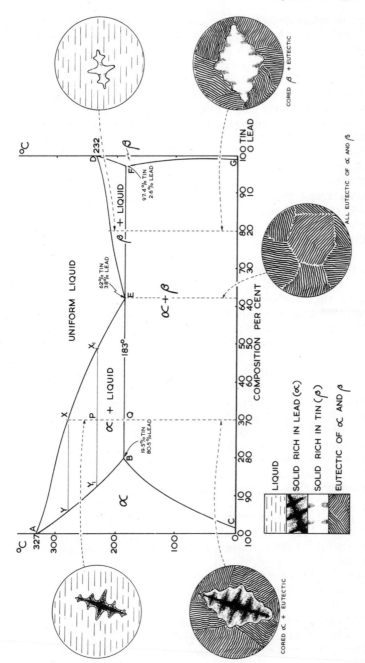

Fig. 9.7—The lead–tin thermal equilibrium diagram, showing the effects of rapid cooling on representative microstructures.

exist, and therefore denote microstructural changes which occur *after an alloy is completely solid*. The slope of *BC* indicates that, as the temperature falls, the solubility of solid tin in solid lead will diminish from 19·5% at 183°C to about 2% at 0°C (point *C*); and similarly the slope of *FG* indicates that the solubility of solid lead in solid tin will fall from 2·6% at 183°C to less than 1% at 0°C (point *G*). Consequently, as our 70% lead/30% tin alloy cools slowly from 183°C to room temperature, the composition of any α in the structure alters along *BC*; whilst the composition of any β will alter along *FG*.

In practice, such an alloy will never cool slowly enough for equilibrium to be reached at each stage of the process, and some coring will inevitably occur, particularly in the crystals of primary solid solution. Accordingly, the sketches representing microstructures in fig. 9.7 assume that cooling has been fairly rapid, and that considerable coring has occurred as a result. Extremely slow cooling (as outlined above) or prolonged annealing eventually eliminates coring.

Precipitation from a solid solution

9.60 In the preceding section, the significance of the sloping solvus lines— *BC* and *FG* (fig. 9.7)—was mentioned. Since it has been the author's experience that many students find ideas of variation in solid solubility difficult to understand, some further space will be devoted to the subject here.

9.61 Let us first consider a parallel case concerning the solubility of a salt in water. If some of the salt is put into water in a beaker, and stirred, much of the salt may dissolve; but some solid may remain at the bottom of the beaker. We thus have two phases in the beaker—a *saturated* solution, and some solid salt. If we now gently warm the solution, more and more salt will dissolve, until only solution remains. At a higher temperature still, the solution would dissolve more salt, if it were available in the beaker. The solution is therefore said to be *unsaturated*, and only a single phase remains in the beaker—the unsaturated solution.

9.62 It is quite easy to plot a curve showing the variation in the solubility of the salt with a rise in temperature. Such a curve is shown in fig. 9.8. It indicates that, as the temperature increases, so does the solubility of salt in water.

Suppose we add some salt to pure water; so that X denotes the total percentage of salt present. After mixing the two together at, say, 10°C, we shall find that we still have a quantity of solid salt remaining. In fact, the solubility diagram tells us that $Y\%$ of salt has actually dissolved, giving a saturated solution at that temperature, and that $(X - Y)\%$ salt remains at the bottom of the beaker. If we now warm the beaker slowly to 30°C, we shall find that more salt dissolves, and the solubility diagram confirms that the amount in solution has increased to Y_1.

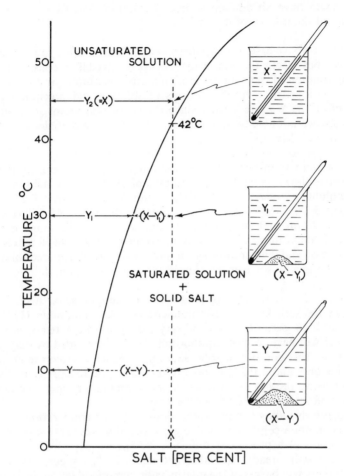

Fig. 9.8—The solubility curve for salt in water.
Solubility increases as the temperature increases.

At, say, 40°C, we would find that a very small quantity of solid salt remained, and at a slightly higher temperature (42°C) this would just dissolve. If the temperature were raised to, say, 45°C, the solution would then be unsaturated—that is, it would dissolve more salt at that temperature, if solid salt were added to the beaker.

9.63 This phenomenon of solution is a reversible process, and, if we allow the beaker to cool slowly, tiny crystals of salt will precipitate when the temperature has fallen a little below 42°C. These crystals will increase in size as the temperature falls; and by the time 10°C has been reached, we shall again have an amount Y left in solution, and $(X - Y)$ as solid crystals at the bottom of the beaker.

9.64 In the above case, we have been dealing with a *liquid* solution of salt in water, but exactly the same principles are involved if we consider instead a *solid* solution of, say, copper in aluminium. Naturally, in the case of a solid solution, the reversible process of solution and precipitation will take place much more slowly, since the individual atoms in a metallic structure are not able to move about as freely as the particles of salt and water in a beaker.

9.65 The aluminium-rich end of the aluminium-copper thermal equilibrium diagram is shown in fig. 9.9. The sloping phase boundary AB shows that the solubility of *solid* copper in *solid* aluminium increases from 0·2% at 0°C (at A) to 5·7% at 548°C (at B). Any point to the left of AB will represent in composition and temperature an unsaturated solid solution (α) of copper in aluminium; whilst any point to the right of AB will represent in composition and temperature a structure consisting of saturated solid solution α, along with some excess aluminium-copper compound.

9.66 We will consider an alloy containing 4% copper, since this forms the basis of the well known aluminium-copper alloy duralumin (17.70). If this has been permitted to cool very slowly to room temperature, its structure will have reached equilibrium, and is represented by diagram (i) (fig. 9.9). This consists of solid solution α, which at room temperature contains only about 0·2% of dissolved copper, the remainder of the 4% copper being present as particles of the aluminium-copper intermetallic compound scattered throughout the structure.

Suppose this alloy is now heated slowly. As the temperature rises, the solid aluminium-copper compound gradually dissolves in the solid solution α, by means of a process of diffusion (8.35). At, say, 300°C, the solid solution α will already have absorbed about 2·2% copper, and for this reason there will be less of the intermetallic compound left in the structure (ii).

At about 460°C (point S), the solution of the intermetallic compound will be complete (iii), the whole of the 4% copper now being dissolved in the solid solution α. In practice, the alloy will be heated to about 500°C (point P), in order to ensure that all of the intermetallic compound has been absorbed by the solid solution α. (Care must be taken not to heat the alloy above the point L, since at this point it would begin to melt.)

The alloy is permitted to remain at 500°C for a short time, so that its solid solution structure can become uniform in composition. It is then quickly removed from the furnace, and immediately quenched in cold water. As a result of this treatment, *the rate of cooling will be so great that particles of the intermetallic compound will have no opportunity to be precipitated.* Therefore, at room temperature we shall have a uniform struc-

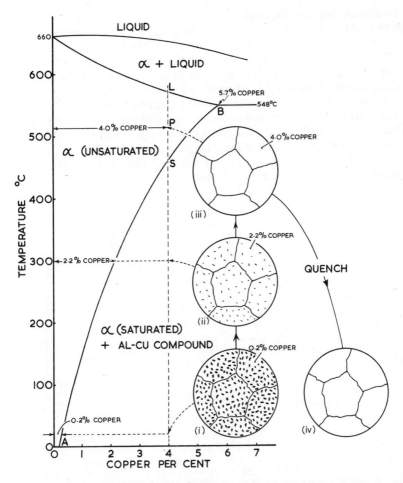

Fig. 9.9—The solubility curve for copper in aluminium.
This is the aluminium-rich end of the aluminium–copper thermal equilibrium diagram.

ture of α solid solution; though normally, and with slow cooling, an equi-librium structure consisting of almost pure aluminium (composition *A*) along with particles of the intermetallic compound would be formed (i). Quenching, however, has prevented equilibrium from being attained. Hence the quenched structure is *not* an equilibrium structure, and is in fact a *super-saturated* solid solution, since α contains much more dissolved copper than is normal at room temperature.

This treatment forms the basis of the first stage in the heat-treatment of duralumin-type alloys, and will be dealt with in detail in Chapter Seventeen.

Ternary equilibrium diagrams

9.70 In this chapter, we have been dealing only with equilibrium diagrams which represent *binary* systems—that is, containing *two* metals. If an alloy contains *three* metals, this will introduce an additional variable quantity for our consideration, since the relative amounts of any two of the three metals can be altered independently. Temperature remains the other 'variable'. Thus we have a system with a total of three variables, and this can be represented graphically only by a three-dimensional system. Such a 'solid' diagram (fig. 9.10) will consist of a base in the form of an equilateral triangle; each point of the triangle representing 100% of one of the three metals (in this case cadmium, tin, and bismuth), whilst ordinates normal to this base represent temperature.

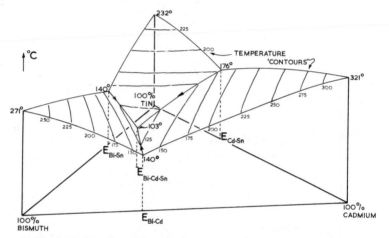

Fig. 9.10—The bismuth–cadmium–tin thermal equilibrium diagram.

This is of course a *ternary* diagram. The three 'valleys' drain down to the ternary (or 'triple') eutectic point at 103°C. This alloy contains 53·9% bismuth, 25·9% tin, and 20·3% cadmium. The 'temperature contours' are at 25°C intervals. The ternary eutectic would melt at a temperature just above that of boiling water.

9.71 A solid diagram of this type is not of much practical use. In the author's student days, research workers used to construct models representing such systems by employing bits of coloured plastic-coated electrical wire to indicate different phase boundaries. With a complex system, the resulting model was quite fantastic, and resembled one of the more lurid examples of modern 'sculpture'—or possibly a parrot cage which had been designed by a committee.

Clearly, if an alloy system contains four or more metals, it will be impossible to represent it by a simple, solid geometrical figure. However, for ternary (that is, three metals) and more complex alloys, we can still draw a useful constitutional diagram by fixing the quantities of all of the metals but one. It is often convenient to do this, as in the case of the diagram for high-speed steel (fig. 13.1). Here we have chosen a high-speed steel of standard composition, and have indicated the effects of variations in the carbon content and the temperature on the structure of this alloy. The amounts of the other alloy additions, viz. tungsten, chromium, and vanadium, are fixed at the values shown. The use of this diagram enables us to explain the basic principles of the heat-treatment of high-speed steel quite adequately. It is not a true equilibrium diagram, but is used like one, and is generally referred to either as a 'constitutional' or a 'pseudo-binary' diagram.

Chapter Ten
Practical Microscopy

10.10 Examination of the microstructure of metals has been practised for little more than a century; yet it is safe to say that, of all the investigational tools available to him, the average metallurgist would least like to be without his microscope. With the aid of quite a modest instrument, a trained metallurgist can obtain an enormous amount of information from the microscopial examination of a metal or alloy. In addition to being able to find evidence of possible causes of failure of a material, he can often estimate its composition, as well as forecast what its mechanical properties are likely to be. Moreover, in the fields of pure metallurgical research, the microscope figures as the most frequently used piece of equipment. The primary object of this chapter is to help the engineering student to acquire some skill in the preparation and examination of a microsection, using a minimum of apparatus.

Selecting a specimen

10.20 Thought and care must be exercised in selecting a specimen from a mass of material, in order to ensure that the piece chosen is representative of the material as a whole. For example, wrought iron contains a considerable amount of slag, which becomes elongated in the direction in which the iron is rolled (fig. 10.1). If only a cross-section A were examined, the

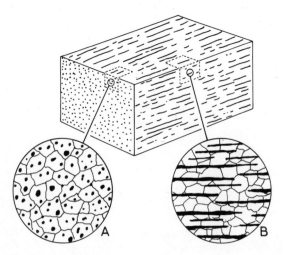

Fig. 10.1—Two specimens are necessary adequately to represent the structure of wrought iron.

observer might be forgiven for assuming that the slag was present in wrought iron as rounded globules, instead of as elongated fibres. The latter fact could only be established if a longitudinal section B were examined in addition to the cross-section A.

10.21 In some materials, both structure and composition may vary across the section. Thus, case-hardened steels have a very different structure in the surface layers from that which is present in the core of the material. The same may be said of steels which have become decarburised at the surface, due to faulty treatment. Frequently it may be necessary to examine two or more specimens in order to obtain comprehensive information on the material.

10.22 A specimen approximately 20 mm in diameter or 20 mm square is convenient to handle. Smaller specimens are best mounted in one of the cold-setting plastic materials available for this purpose, since such a specimen may rock during the grinding process, giving rise to a bevelled surface. Moreover, mounting in plastic affords a convenient way of protecting the edge of a specimen in cases where investigation of the edge is necessary, as, for example, in examining a section through a carburised surface.

One such plastic material is sold under the trade name of NHP.* It

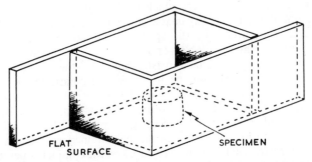

Fig. 10.2—A method of mounting a specimen in a cold-setting plastic material. No pressure is required.

consists of a white powder which, when wetted with the liquid supplied along with it, hardens to give a solid plastic substance which will retain the metal specimen during and after the polishing operation. A simple 'mould' of the type shown in fig. 10.2 is all that is required to mount specimens in this material. The specimen is placed face-down on a sheet of glass which has first been lightly smeared with vaseline. The two L-shaped retaining pieces are then placed around the specimen, as shown, to give a mould of convenient size. The specimen is then covered with powder,

* NHP mounting plastic is an acrylic resin (23.67) supplied by Messrs North Hill Plastics Ltd, 49 Grayling Rd, London N16.

which is in turn moistened with the liquid supplied. If a surplus of liquid is accidentally used, this can be absorbed by sprinkling a little more powder on the surface. In about thirty minutes the mass will have hardened, and the L-shaped members can be detached.

Grinding and polishing the specimen

10.30 It is first necessary to obtain a flat surface on the specimen. This is best achieved by using a file held flat on the work-bench, and then rubbing the specimen on the file. It is much easier for an unskilled operator to produce a single flat surface on the specimen by using this technique, rather than by using a file in the orthodox manner. When the original hack-saw marks have been eliminated, the specimen should be rinsed in running water, to remove any coarse grit which may otherwise be carried over to the grinding papers.

10.31 Grinding is then carried out by using emery papers of successively finer grades. These papers must be of the best quality, particularly in respect of uniformity of particle size. For successful wet-grinding, at least four grades of paper are required ('220', '320', '400', and '600', from coarse to fine), and these must be of the type with a waterproof base. Special grinding tables can be purchased, in which the standard 300 mm × 50 mm strips of grinding papers can be clamped. The surface of the papers is flushed by a current of water, which serves not only as a lubricant in grinding, but also carries away coarse emery particles, which might otherwise scratch the surface of the specimen. If commercially produced grinding tables are not available—and certainly the prices of these simple pieces of apparatus seem to be unreasonably high—there is no reason why simple equipment should not be improvised, as indicated in fig. 10.3. Here a sheet of 6 mm plate glass about 250 mm × 100 mm has a sheet of paper clamped to its surface by a pair of stout paper-clips. The paper should be folded round the edge of the glass plate, so that it will be held firmly. A suitable stream of water can be obtained by using a piece of rubber tubing attached to an ordinary tap, and the complete operation may be conducted in the laboratory sink. Alternatively, the apparatus can be contained in an old photographic developing dish, fitted with a suitable drain, in order to carry away the stream of water. The glass plate is tilted at one end, so that the water flows fairly rapidly over the grinding paper.

10.32 The specimen is first ground on the '220' grade paper. This is achieved by rubbing it back and forth on the paper, in a direction which is roughly *at right angles* to the scratches left by the filing operation. In this way, it can easily be seen when the original deep scratches left by the file have been completely removed. If the specimen were ground so that the new scratches ran in the same direction as the old ones, it would be virtually impossible to see when the latter had been erased. With the primary grinding marks removed, the specimen is now washed free of '220' grit.

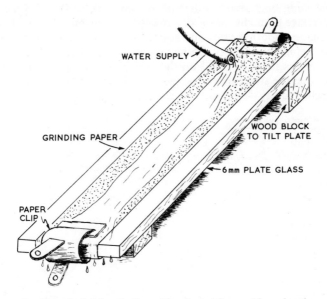

WATER SUPPLY

GRINDING PAPER

WOOD BLOCK
TO TILT PLATE

6mm PLATE GLASS

PAPER
CLIP

Fig. 10.3—A simple grinding table adapted from odds and ends.

Grinding is then continued on the '320' paper; again turning the specimen through 90°, and grinding until the previous scratch marks have been erased. This process is repeated with the '400' and '600' papers.

If circumstances demand that dry-grinding be used, complete cleanliness must be maintained at all stages, in order to avoid the carrying-over of coarse grit to the finer papers. After use, each paper should be shaken free of grit by smartly pulling it taut a number of times. Papers can be stored safely between the pages of a *glossy* magazine, such as *Tatler* (or *Men Only*, for that matter). Alternatively, a strip of each grade of paper can be permanently attached to its own polishing block. It is most important that the specimen be washed before passing from one grade of paper to the next, and particularly before transferring to the final polishing cloth.

Steels and the harder non-ferrous metals can be ground dry, provided that care is taken not to over-heat them, since this may modify the microstructure. For the softer non-ferrous materials, such as aluminium alloys, and bearing metals, the paper should be moistened with a lubricant such as paraffin. A lighter pressure can then be used, and there will be much less risk of particles of grit becoming embedded in the soft metal surface. Modern wet-grinding processes are far more satisfactory for all materials, and have generally replaced dry-grinding methods.

10.33 Up to this stage, the process has been one of grinding, and each set of parallel 'furrows' has been replaced successively by a finer set. The final polishing operation is different in character, in that it removes the ridged surface layers by means of a burnishing process. Although the surface is made smooth by this operation, the structure still cannot be seen, because the nature of the polishing process is such that it leaves a 'flowed' or amorphous layer of metal on the surface of the specimen (fig. 10.4). In order that the structure can be seen, this flowed layer must be dissolved —or 'etched' away—by a suitable chemical reagent.

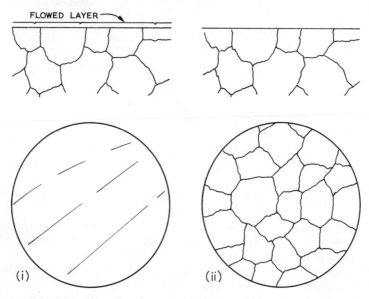

Fig. 10.4—The 'flowed layer' on the surface of a polished microsection.
(i) In the polished state the structure is hidden by the flowed layer—only a few polishing scratches are visible on the otherwise featureless surface.
(ii) Etching removes the flowed layer, thus revealing the crystal structure beneath.

10.34 Iron or steel specimens are polished by means of a rotating cloth pad impregnated with a suitable polishing medium. 'Selvyt' cloth is probably the best-known material with which to cover the polishing wheel, though special cloths are now available for this purpose. The cloth is thoroughly wetted with *distilled* water, and a small quantity of the polishing powder is gently rubbed in with *clean* finger-tips. Possibly the most popular polishing medium is alumina (aluminium oxide), generally sold under the name of 'Gamma Alumina'. During the polishing process, water should be permitted to spot on to the pad, which should be run at a low speed until the operator has acquired the necessary manipulative technique. Light pressure should be used, since too heavy a pressure on the specimen may result in a torn polishing cloth. Moreover, the specimen is more likely to be scratched by grit particles embedded deep in the cloth if heavy pressure is applied.

A disadvantage of most of the water-lubricated polishing powders is that they tend to dry on the pad, which generally becomes hard and gritty as a result. If the pad is to be used only intermittently, it might be worthwhile to use one of the proprietary diamond-dust polishing compounds. In these materials, graded diamond particles are carried in a 'cream' base which is soluble both in water and in the special polishing fluid, a few spots of which are applied to the pad in order to lubricate the work and lead to even spreading of the polishing compound. If properly treated, the pad remains in good condition until it wears out; so, although these diamond dust materials are more expensive than other polishing media, a saving may result in the long term, as polishing cloths will need to be changed less frequently.

10.35 Since non-ferrous metals and alloys are much softer than steels, they are best polished by hand, on a small piece of 'Selvyt' cloth wetted with 'Silvo'. During polishing, a circular sweep of the hand should be used, rather than the back-and-forth motion used in grinding.

When the surface appears free from scratches, it is cleaned thoroughly, dried, and then examined under the microscope, using a magnification between 50 and 100. If reasonably free from scratches, the specimen can at this stage be examined for inclusions, such as those of manganese sulphide (in steel), slag fibres (in wrought iron), or globules of lead (in free-cutting brasses). Such inclusions would be less obvious were the specimen etched *before* this primary examination.

10.36 The most important points to be observed during the grinding and polishing processes are:

(1) Absolute cleanliness is necessary at each stage.

(2) Use very light pressure during both grinding and polishing.

(3) Deep scratches are often produced during the final stage of grinding. Do *not* attempt to remove these by prolonged polishing, as such scratches tend to be obliterated by the flowed layer, only to reappear on etching. Moreover, prolonged polishing of non-ferrous metals tends to produce a rippled surface. If deep scratches are formed, wash the specimen, and return to the last-but-one paper, remembering to grind in a direction at 90° to the scratches.

(4) Care should be taken not to overheat the specimen during preliminary filing or grinding. Hardened steels could be tempered by such treatment, particularly if a linishing machine is used.

Etching the specimen
10.40 Etching is generally the stage in preparing a microsection that the beginner finds most difficult. Often the first attempt at etching results in a badly stained or discoloured surface, and this is invariably due to inadequate cleaning and degreasing of the specimen before attempting to etch it.

The specimen should first be washed free of any adhering polishing compound. This can be rubbed from the *sides* of the specimen using the fingers, but great care must be taken in dealing with the polished face. The latter can be cleaned and degreased successfully by *very gently* smearing the surface with a *clean* finger-tip dipped in grit-free soap solution, followed by thorough rinsing under the tap. Even now, traces of grease may still be present; as shown by the fact that a film of water will not flow evenly over the surface, but instead remains as isolated droplets. The last traces of grease are best removed by immersing the specimen for a minute or so in boiling* alcohol ('white industrial methylated spirit').

From this stage onwards, the specimen should not be touched by the fingers, but be handled with a pair of nickel tongs. It is lifted from the alcohol, and cooled under the tap before being etched. Some thermoplastic mounting materials are dissolved by hot alcohol; in such cases, swabbing with a piece of cotton wool saturated with dilute caustic-soda solution may degrease the surface effectively.

10.41 When the specimen is clean and free of grease, it is etched by plunging it into the etching solution, and agitating it vigorously for a few seconds. The specimen is then *very quickly* transferred to running water, in order to wash away the etchant as rapidly and as evenly as possible. It is then examined with the naked eye, to see to what extent etching has taken place. If successfully etched, the highly polished surface will now appear dull, and, in the case of cast metals, individual crystals may be seen. A bright surface at this stage will usually indicate that further etching is necessary. The time required for etching varies with different alloys and etchants, and may be limited to a few seconds for a specimen of carbon steel etched in 2% 'nital', or extend to as long as 30 minutes for a stainless steel etched in a mixture of concentrated acids.

After being etched, the specimen is washed in running water, and then quickly immersed in boiling alcohol, where it should remain for a minute. On withdrawal from the alchol, the specimen is shaken with a flick of the tongs, to remove surplus alcohol so that it will dry almost instantaneously. With specimens mounted in a plastic likely to be affected by boiling alcohol, it is better to spot a few drops of cold alchol on the surface of the specimen. The surplus is then shaken off, and the specimen is held in a current of hot air from a domestic hair-drier. Unless the specimen is dried *evenly and quickly*, it will stain.

A summary of the more popular etching reagents which can be used for most metals and alloys is given in table 10.1.

---

* On no account should alcohol be heated over a naked flame, as the vapour is highly inflammable. An electrically heated water-bath should be used—an electric kettle with the lid removed is serviceable.

Type of etchant	Composition	Characteristics and uses
2% 'nital'—for iron, steel, and bearing metals	2 cm³ nitric acid, 98 cm³ alcohol ('white industrial methylated spirit')	The best general etching reagent for irons and steels, both in the normalised and heat-treated conditions. For pure iron and wrought iron, the quantity of nitric acid may be raised to 5 cm³. Also suitable for most cast irons and for some alloys, such as bearing metals.
Alkaline sodium picrate—for steels	2 g picric acid, 25 g sodium hydroxide, 100 cm³ water	The sodium hydroxide is dissolved in the water, and the picric acid is then added. The whole is heated on a boiling water-bath for 30 min, and the clear liquid is poured off. The specimen is etched for 5–15 min in the boiling solution. It is useful for distinguishing between ferrite and cementite; the latter is stained black, but ferrite is not attacked.
Ammonia/ hydrogen peroxide—for copper, brasses, and bronzes	50 cm³ ammonium hydroxide (0·880), 50 cm³ water. Before use, add 20–50 cm³ hydrogen peroxide (3%).	The best general etchant for copper, brasses, and bronzes. Used for swabbing or immersion. Must be freshly made, as the hydrogen peroxide decomposes. (The 50% ammonium hydroxide solution can be stored, however.)
Acid ferric chloride—for copper alloys	10 g ferric chloride, 30 cm³ hydrochloric acid, 120 cm³ water	Produces a very contrasty etch on brasses and bronzes. Use at full strength for nickel-rich copper alloys, but dilute one part with two parts of water for brasses and bronzes.
Dilute hydrofluoric acid—for aluminium and its alloys	0·5 cm³ hydrofluoric acid, 99·5 cm³ water	A good general etchant for aluminium and most of its alloys. The specimen is best swabbed with cotton-wool soaked in the etchant. N.B. *On no account should hydrofluoric acid or its fumes be allowed to come into contact with the eyes or skin. Care must be taken with all concentrated acids.*

Table 10.1—Etching reagents.

Details of etching reagents for other specific purposes will be found in *Engineering Metallurgy, Part 1* (section 10.24), by this author.

The metallurgical microscope

10.50 The reader may have used a microscope during his school days, but the chances are that this would be an instrument designed for biological work. Biological specimens can generally be prepared as thin, transparent slices, mounted between thin sheets of glass, so that illumination can be arranged simply by placing a source of light *behind* the specimen. Since metals are opaque substances, which must be illuminated by frontal lighting, the source of light must be *inside* the microscope tube itself. This is generally accomplished as shown in fig. 10.5, by placing a small,

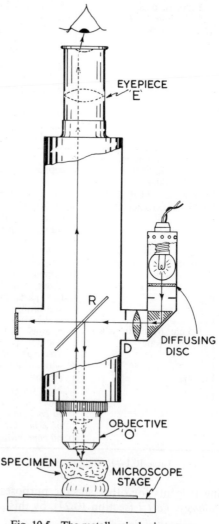

Fig. 10.5—The metallurgical microscope.

thin plain-glass reflector, R, inside the tube. Since it is necessary for the returning light to pass through R, the latter must be of unsilvered glass. This means that much of the total light available is lost, both by transmission through the glass when it first strikes the plate, and by reflection when the returning ray from the specimen strikes the plate again. Nevertheless, a small 6-volt bulb is generally sufficient as a source of illumination. The width of the beam is controlled by the iris diaphragm, D. This should be closed until the width of the beam of light is just sufficient to cover the rear component of the objective lens, O. Excess light, reflected within the microscope tube, would cause light-scatter and consequently 'glare' in the field of view, leading to a loss of contrast and definition in the image formed.

10.51 The optical system of the microscope consists of two lenses: the objective, O, and the eyepiece, E. The former is the more important and expensive of the two lenses, since it must resolve fine detail of the object under examination. Good-quality objectives, like camera lenses, must be of compound construction. However, there is a limit to the degree of accuracy which is worthwhile in constructing an objective. At magnifications of ×1000 or so, one is dealing with dimensions comparable with the wavelength of light itself, and further improvements in the quality of the lens would produce no corresponding improvements in the 'sharpness' of the image. Thus, in purchasing a very expensive microscope, one may well be paying for extra refinements which may not be necessary. It is doubtful whether one will obtain higher optical quality than is offered in the 'standard' model of the same manufacturer's range of products. The same is not true of ordinary camera lenses, however; here, as a general rule, the more one is prepared to pay, the better the quality of the lens.

10.52 The magnification given by an objective depends upon its focal length—the shorter the focal length, the higher the magnification obtainable. As mentioned above, the resolving power of a lens is also important, and, within the limits stated, depends upon the quality of the lens. When working at high magnifications of ×1000 or more, it is generally useless to increase the size of the image either by extending the tube length or by using a higher-power eyepiece, as a point is reached where there is a falling-off in definition. A parallel example in photography is where an enlargement of a small 9 cm × 6 cm snapshot fails to show any more *detail* than it did in its original size, but instead gives a rather blurred image.

10.53 The eyepiece is so called because it is the lens nearest the eye. Its purpose is to magnify the image formed by the objective. Eyepieces are made in a number of powers, generally ×6, ×8, ×10, and ×15.

10.54 The distance separating the objective and the eyepiece is known as the tube-length of the microscope, and is usually 200 mm for most instruments. The magnification of the system can then be calculated from:

$$\text{magnification} = \frac{\text{tube-length (mm)} \times \text{power of eyepiece}}{\text{focal length of objective (mm)}}$$

Thus, for a microscope having a tube-length of 200 mm, and using a 4 mm focal-length objective in conjunction with a ×8 eyepiece,

$$\text{magnification} = \frac{200 \times 8}{4}$$

$$= 400$$

Using the microscope

10.60 The specimen must first be mounted so that its surface is normal to the axis of the instrument. This is best achieved by fixing the specimen to a microscope slide, by means of a pellet of 'plasticine', using a mounting-ring to ensure normality between the surface of the specimen and the axis of the microscope (fig. 10.6). Obviously the mounting-ring must have perfectly parallel end-faces.

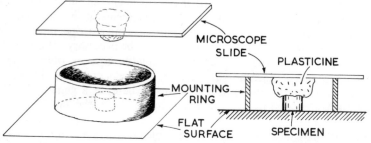

Fig. 10.6—Mounting a specimen so that its surface is normal to the axis of the microscope.

10.61 The specimen is brought into focus by first using the coarse adjustment, and then the fine adjustment. Lenses are designed to work at an optimum tube-length (usually 200 mm), and give best results under these conditions. Hence the tube carrying the eyepiece should be drawn out to the appropriate mark (a scale is generally engraved on the side of the tube). Slight adjustments in tube-length should then be made to suit the individual eye. Finally, the iris, D, in the illumination system should be closed to the point where illumination begins to decrease. Glare due to internal reflection will then be at a minimum.

10.62 Invariably, the newcomer to the microscope selects the combination of objective and eyepiece which will give him the maximum magnification, but it is a mistake to assume that high magnifications in the region of ×1000 are necessarily the most useful. In fact, they may well give a misleading impression of the structure, by pin-pointing some very localised feature rather than giving a general picture of the microstructure of the material. Directional properties in wrought structures, or dendritic formations in cast structures, are best examined using low powers of between ×30 and ×100. Even at ×30, a single crystal of a cast structure may completely fill the field of view. The dendritic pattern, however, will be clearly apparent; whereas at ×500, only a small area between two dendrite arms would fill the field of view. In a similar way, a more representative impression of the lunar landscape may be obtained by using a pair of good binoculars than by using the very high-powered system of an astronomical telescope. Hence, as a matter of routine, a low-powered objective should *always* be used in the initial examination of a microstructure, before it is examined at a high magnification.

The care of the microscope
10.70 Care should be taken never to touch the surface of optical glass with the fingers, since even the most careful cleaning may damage the surface coating (most high-quality lenses are 'bloomed'; that is, coated with magnesium fluoride to increase light transmission). In normal use, dust particles may settle on a lens, and these are best removed by sweeping gently with a high-quality camel-hair brush.

10.71 If a lens becomes finger-marked, this is best dealt with by wiping *gently* with a piece of soft, well washed linen moistened with the solvent xylol. Note that the operative word is *wipe*, not *rub*. Excess xylol should be avoided, as it may penetrate into the mount of the lens, and soften the cement holding the glass components together.

High-power objectives of the oil-immersion type should be wiped free of cedar-wood oil before the latter has a chance to harden. If, due to careless neglect, the oil has hardened on a lens, then it will need to be removed with the minimum amount of xylol; but the use of the latter should be avoided whenever possible.

Soft, well washed linen should always be used to clean lenses. It is superior to chamois leather, which is likely to absorb particles of grit, and to silk, which has a tendency to scratch the surface of soft optical glass.

The electron microscope
10.80 As the reader will have gathered, the greater proportion of routine microscopic examination of metals is carried out using magnifications in the region of ×100. It is often necessary, particularly in the field of research, to examine structures at much higher magnifications. Unfortunately, it is not practicable to use magnifications in excess of ×2000 with an ordinary optical microscope, since, as indicated earlier in this chapter (10.51), one is then dealing with an object size of roughly the same order as the wavelength of light itself. This leads to a loss in definition of the image at magnifications greater than ×2000.

For high-power microscopy—between ×2000 and about ×200 000—

an electron microscope may be used. In this instrument, rays of light are replaced by a beam of electrons which have been made to travel at a suitable velocity. Since electrons are very tiny* electrically charged particles, their path can be altered by an electromagnetic field. Consequently, the 'lenses' in an electron microscope consist of a system of coils which produce the necessary electromagnetic field to focus the electron beam. The other important difference between an optical and an electron microscope is that, whereas in the former light is *reflected* from the surface of the specimen, in the electron microscope the electron beam *passes through* the specimen, rather in the manner of light rays in a biological microscope. Hence, thin foil specimens must be used, or, alternatively, a very thin replica of the etched metallic surface may be produced in a suitable plastic material. This replica is then examined by the electron microscope.

10.81 The limitations of the optical microscope have already been mentioned, but it should be noted that the range of the ordinary electron microscope too is limited. Whilst it is true that magnifications obtained with this instrument exceed a quarter of a million times, it should be appreciated that a magnification of something like twenty million times is required for us to be able to see an average-size atom. Magnifications of this order can now be obtained using a modern *field-ion* microscope, and the results so far obtained at least confirm the physicist's idea of the general form of atoms.

Sketching microstructures
10.90 It is inevitable that at some stage in his education the student will be required to sketch microstructures of metals and alloys. In the writer's experience, most attempts at this turn out as dismal failures, and he is continually marking examination papers in which the sketch of an alleged microstructure resembles that of a dry-stone wall more than anything else. Although metallurgical structures are generally illustrated by photomicrographs, it is useful to be able to make a reasonably accurate sketch of such a structure. Moreover, a simplified sketch will often illustrate a particular point more effectively than will a more complex photomicrograph.

Two things are essential in making a successful sketch of a microstructure: a sharp pencil, and plenty of patience. A 'B' pencil is possibly the best type to use, though the individual may prefer 'HB' (or even 'H', if he is a draughtsman). It is best sharpened to a fine point using a pencil sharpener.

10.91 The sketch is most conveniently contained in a circle of about 65 mm diameter; but, before starting to make the sketch, it is necessary to study the structure carefully for some time, in order to decide which are the most

* The mass of the electron is $9\cdot11 \times 10^{-31}$ kg; that is, about 1/1840 of the size of the smallest atom—that of hydrogen.

important features to be depicted. Often the view through the lens may disclose a bewildering amount of detail, and a higher magnification may then simplify the problem by concentrating attention on a smaller area of the structure, and at the same time make the sketching of a mass of very fine detail unnecessary. Whatever magnification is chosen, it is necessary that the field of view adequately represents the structure as a whole.

From then on it is a matter of patience. Frequent glances through the microscope are followed by the transfer of a few lines to the sketch, until ultimately the picture begins to resemble that which is observed through the lens. Patience is essential, though training and the ability to sketch are of course useful attributes. If carefully executed, a sketch can show the essential features of a structure more clearly than can a photomicrograph, and in this connection the reader's attention is directed to the sketches of the microstructures of cast irons included in this book (figs 15.1 and 15.4).

Chapter Eleven
Iron and Steel

11.10 Since the onset of the Industrial Revolution, the material wealth and power of a nation has depended largely upon its ability to make steel. During the nineteenth century, Britain was prominent among steel-producing nations, and, towards the end of the Victorian era, was manufacturing a great proportion of the world's steel. Exploitation of vast deposits of ore abroad has changed the international situation during the present century; so that now the two 'super-powers'—the United States of America and the USSR—owe their material power largely to the presence of high-grade iron ore within, or very near to, their own vast territories.

We must not lose sight of the fact, however, that the output of steel from British furnaces has not actually fallen. Indeed, it is higher now than it was in the great days of the British Empire: it is just that we have been overtaken by other countries better placed than we are to increase their output of steel. Until recently, we ran third in the international steel-making stakes, but we have of recent years been outflanked by both West Germany and Japan.

Fortunately for us, most of the rather low-grade ore available in Britain is near to the surface; so it can be mined by relatively cheap open-cast methods. Nevertheless, we have to import high-grade ore concentrates from abroad—particularly from Sweden and North Africa—in order to make up the deficiency in home-produced ore.

11.11 Smelting of iron ore takes place in the blast-furnace (fig. 11.1). A modern blast-furnace is something like 60 m high and 7·5 m in diameter at the base, and may produce from 2000 to 10000 tonnes of iron per day. Since a refractory lining lasts for several *years*, it is only at the end of this period that the blast-furnace is shut down; otherwise it works a 365-day year. Processed ore, coke, and limestone are charged to the furnace through the double-bell gas-trap system; whilst a blast of heated air is blown in through the tuyères near the hearth of the furnace. At intervals of several hours, the furnace team opens both the slag hole and the tap hole, in order to run off first the slag, and then the molten iron. The holes are then plugged with clay.

11.12 The smelting operation involves two main reactions.

(1) The chemical reduction of iron ore by carbon monoxide gas arising from the burning coke:

iron oxide + carbon monoxide → iron + carbon dioxide
(in ore)

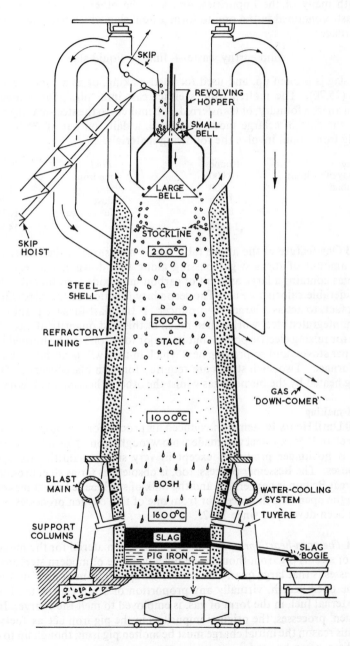

Fig. 11.1—A modern blast-furnace.

(2) Lime (from limestone added with the furnace charge) combines with many of the impurities, and also the otherwise infusible earthy waste contained in the ore, to form a fluid slag which will run from the furnace:

$$\text{solid earthy waste} + \text{lime} \rightarrow \text{liquid slag}$$

The slag is broken up, and used for road-making, or as a concrete aggregate (25.70). The molten iron is either cast into 'pigs', for subsequent use in an iron foundry, or transferred, still molten, to the steel-making plant.

In the case of a large modern furnace, a daily output of 2000 tonnes of pig iron would involve the following materials.

Charge	(tonnes)	Products	(tonnes)
ore (say 50% iron)	4 000	pig iron	2 000
limestone	800	slag	1 600
coke	1 800	dust	200
air	8 000	furnace gas	10 800
	14 600		14 600

11.13 One feature of the above table which may surprise the reader is the vast amount of furnace gas passing along the 'down-comer' each day. The gas contains a large amount of carbon monoxide, and therefore has a considerable calorific value. The secondary function of the blast-furnace is, in fact, to act as a large gas-producer. If the blast-furnace plant is part of an integrated steel works, then much of this vast quantity of gas will be used for raising electric power; but its major function is to be burned in the Cowper stoves, and so provide heat which in turn will heat the air blast to the furnace. Two such stoves are required for each blast-furnace. One is being heated by the burning gas whilst the other is heating the ingoing air.

Steel-making

11.20 Until Henry Bessemer introduced his process for the mass-production of steel, in 1856, all steel was made from wrought iron. Nowadays wrought iron is no longer produced, except in very small quantities for special purposes. The Bessemer process, too, is obsolete as far as steel production in Great Britain is concerned, and the bulk of steel made here comes either from the open-hearth process or from one of the 'oxygen processes' which have been developed since 1952.

11.21 *The open-hearth process* The principal raw material for the manufacture of steel is pig iron, though one big advantage of modern steel-making processes is that large quantities of steel scrap can be used up if necessary. In the open-hearth, virtually any proportion of scrap can be used, since an external fuel, in the form of gas, is employed to melt the charge. In the 'oxygen' processes, the original impurities in the pig iron act as fuel; and for this reason the initial charge must be molten pig iron, though up to 40% solid scrap can be used.

The main features of the open-hearth furnace are represented diagramatically in fig. 11.2. As its name suggests, this furnace consists mainly of a large open hearth, which may hold up to 100 tonnes or more of molten steel.

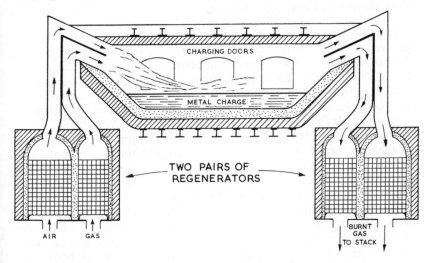

Fig. 11.2—The open-hearth furnace for making steel.

Since a very high temperature (1600°C) is required in this process, the ingoing gas and air need to be pre-heated. This is achieved by using two pairs of 'regenerators'. These are chambers containing firebrick assembled honeycomb-fashion, and through which gas or air can pass. One pair— that on the right of the diagram—is being reheated by the outgoing burnt gas, whilst the other pair—that on the left—having already been heated in this way, is now being used to heat the ingoing air and gas. The flow of gases will be reversed when the pair of stoves on the right has reached the necessary high temperature. Cowper stoves are employed in a similar manner, to heat the air supply to the blast-furnace.

The process of steel-making is mainly one involving oxidation of the impurities present in the initial charge; so that they form a slag which floats on the surface of the molten steel. The oxidation process is achieved partly by air admitted to the furnace chamber, and partly by high-grade oxide ore added to the charge. Most British steel-making is of the 'basic' type; that is, lime is added to the charge, in order to form a basic slag, and so remove the phosphorus which is present in most British pig irons. Basic slag is a valuable by-product, used as an agricultural fertiliser.

When the impurities have been reduced to the necessary low level, the slag is run off; and carbon is added in the form of anthracite, much of which dissolves in the molten steel. Finally, the molten steel is transferred to a large ladle, which in turn pours its charge into cast-iron moulds, to produce ingots, each of several tonnes in mass. The main advantage of the basic open-hearth process is that it can utilise almost any grade of raw material, and turn it into high-quality steel. Nevertheless the open-hearth is destined to follow the Bessemer process into obscurity as BOS (Basic Oxygen Steelmaking) takes over.

11.22 *Basic oxygen steelmaking.* In the Bessemer process, impurities were removed from the charge of molten pig iron by blowing air through it. The impurities, mainly carbon, phosphorus, silicon, and manganese, acted as fuel; and so the range of compositions of pig iron was limited, because sufficient impurities were necessary in order that the charge did not 'blow cold' from lack of fuel. The oxidised impurities either volatilised or formed a slag on the surface of the charge.

Since the air blast contained only 20% of oxygen by volume, much valuable heat was carried away from the charge by the 80% nitrogen also present. Worse still, a small amount of this nitrogen dissolved in the charge, and, in the case of mild steel destined for deep-drawing operations, caused a deterioration in its mechanical properties. The new oxygen processes produce mild steel very low in nitrogen; so that its deep-drawing properties are superior to those of the old Bessemer steel. Improvements of this type are essential if mild steel is to survive the challenge of re-inforced plastics such as ABS (23.628) in the field of automobile bodywork.

11.23 One of the leading oxygen processes is the L-D process, so called because it originated in the Austrian industrial towns of Linz and Donawitz, in 1952. The L-D converter (fig. 11.3) is a pear-shaped vessel of up to 300 tonnes capacity. Since, in this process, no heat is carried away by useless nitrogen, a charge containing up to 40% scrap can be used. This scrap is loaded to the converter first, followed by lime and molten pig iron.

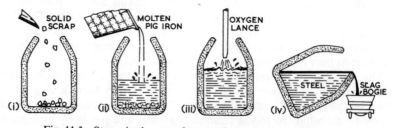

Fig. 11.3—Stages in the manufacture of steel by the L-D process.
Steel scrap is added first (i), followed by molten pig iron (ii). At the end of the 'blow' (iii), the slag is run off first (iv), before 'teeming' the steel into a ladle.

Oxygen is then blown at the surface of the molten charge from a water-cooled 'lance', which is lowered through the mouth of the converter to within 0·5 m of the surface of the metal. Impurities in the charge are oxidised, and form a slag on the surface. At the end of the 'blow', this slag is run off first; and the charge is transferred to a ladle, preparatory to being cast as ingots.

The composition of steel

11.30 Plain-carbon steels are those alloys of iron and carbon which contain up to 1·7% carbon. In practice, most ordinary steels also contain up to 1·0% manganese, which is left over from a deoxidation operation carried out at the end of the steel-making process. The presence of this excess of manganese helps to reduce the sulphur content of the steel. Both sulphur and phosphorus are extremely harmful impurities which give rise to brittleness in steels. Consequently, most specifications allow no more than 0·06% of either of these elements; whilst specifications for higher-quality steels limit the amount of each element to 0·04%. The manganese mentioned above dissolves in the solid steel, slightly increasing its strength and hardness.

11.31 At ordinary temperatures, most of the carbon in a steel which has not been heat-treated is chemically combined with some of the iron, forming an extremely hard compound known by chemists as iron carbide,* though metallurgists often refer to it as 'cementite'. Since cementite is very hard, the hardness of ordinary carbon steel increases with the carbon content. Carbon steels can be classified into groups, as shown in fig. 11.6 and in table 12.2.

The structure of plain-carbon steels

11.40 In Chapter 2 (2.14) we saw that iron is what is called an allotropic element; that is, an element which leads a sort of Jekyll-and-Hyde existence by appearing in more than one physical form. Below 910°C, pure iron has a body-centred cubic crystal structure; but on heating the metal to a temperature above 910°C, its structure changes to one which is face-centred cubic. Now face-centred cubic iron will take quite a lot of carbon—up to 1·7% in fact—into solid solution (8.34); whereas body-centred cubic iron will dissolve scarcely any—a maximum of only 0·03%. Since the solid solubility of carbon in iron alters in this way, it follows that changes in the structure will also occur on heating and cooling. Thus it is the allotropic transformation, and the structural changes which accompany it, which cause the thermal equilibrium diagram (fig. 11.4) to have a somewhat unusual shape as compared with those already dealt with in Chapter 9.

11.41 Any solid solution of carbon up to a maximum of 1·7% in face-centred cubic iron is called *austenite* (γ); whilst the very dilute solid solution formed when up to 0·03% carbon dissolves in body-centred cubic iron is called *ferrite* (α). For all practical purposes, we can regard ferrite as being more or less pure iron, since less than 0·03% carbon will have little effect on its properties. Thus, in a carbon steel at, say, 1000°C, all of the carbon present is dissolved in the solid austenite. When this steel cools,

---

* For those readers who have a knowledge of chemistry, the chemical formula for iron carbide is Fe_3C.

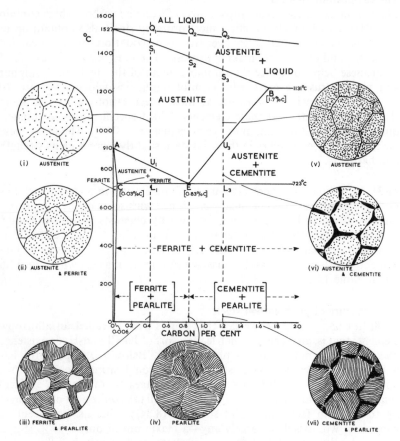

Fig. 11.4—The iron-carbon equilibrium diagram.
The small dots in the diagrams representing structures containing austenite do *not* represent visible particles of cementite—they are meant to show the actual concentration of carbon *dissolved* in the austenite, and in the real structure they would be invisible.

the austenite changes to ferrite, which will retain practically no carbon in solid solution. 'What happens to this carbon?', we may ask. The answer is that, assuming the cooling has taken place fairly slowly, the carbon will be precipitated as the hard compound *cementite*, referred to previously (11.31).

11.42 By referring to fig. 11.4, let us consider what happens in the case of a steel containing 0·4% carbon as it solidifies and cools to room temperature. It will begin to solidify at a temperature of about 1500°C (Q_1), and will be completely solid by the time the temperature has fallen to about 1450°C (S_1). The structure at this stage will be of uniform austenite. As this uniform austenite cools, nothing further will happen to its structure—except possibly grain growth—until it reaches the point U_1, which is known as the 'upper critical point' for this particular steel. Here austenite begins to change to ferrite, which will generally form as small new crystals at the grain boundaries of the austenite (fig. 11.4 (ii)). Since ferrite contains very little carbon, it follows that at this stage the bulk of the carbon must remain behind in the shrinking crystals of austenite; and so the composition of the latter moves to the right. Thus, by the time the temperature has fallen to 723°C, we shall have a mixture of ferrite and austenite crystals of compositions C and E respectively. The *overall* composition of the piece of steel is given by L_1, and so we can apply the lever rule:

weight of ferrite (composition C) . $CL_1 =$
$$\text{weight of austenite (composition } E \text{) } . L_1E$$

Since CL_1 and L_1E are of more or less equal length, it follows that the amounts of ferrite and austenite at this temperature (723°C) are roughly equal for this particular composition of steel (0·4% carbon).

The reader will recognise the point E as being similar to the eutectic points dealt with in Chapter 9 (figs 9.6 and 9.7). In the present case, however, we are dealing with the transformation of a *solid* solution (austenite), instead of the solidification of a *liquid* solution. For this reason, we refer to E as a eutect*oid* point, instead of a eutect*ic* point.

11.43 As the temperature falls just below 723°C (the 'lower critical temperature'), the austenite, now of composition E, transforms to a eutectoid (fig. 11.4 (iii)) by forming alternate layers of ferrite (composition C) and the compound cementite (containing 6·9% carbon). Clearly, since the austenite at this temperature was of composition E (0·83% carbon), the *overall* composition of the eutectoid which forms from it will be of composition E (0·83% carbon), even though the separate layers comprising it contain 0·03% and 6·9% carbon respectively.

As the steel cools down to room temperature, no further important change will take place in the structure.

11.44 The mechanism of the above transformation of austenite reminds the author of a travel film he saw some time ago, in which the behaviour of a group of penguins on an ice-floe struck him as being similar to the behaviour of the carbon atoms in austenite. As the ice-floe melted and changed to water (austenite turns to ferrite), the penguins (cementite), not wishing to get wet, huddled closer and closer together on the floe, until a point was reached when there was not room for all of them (austenite is saturated with carbon). Consequently, they were pushed, one by one, into the water as the ice melted away. Just as the final result in this analogy is a mixture of penguins and water, so is the eutectoid mentioned here a mixture of ferrite and cementite.

Fig. 11.5—The analogy between penguins on an ice-floe and the carbon atoms in austenite.

11.45 In most cases, a eutectic or eutectoid in an alloy system is not given a separate name, since it is really a mixture of two phases. The iron-carbon system, however, is the most important of all alloy systems with which the metallurgist or engineer has to deal; so the eutectoid of ferrite and cementite referred to above is given the special name of *pearlite*. This name is derived from the fact that the etched surface of a high-carbon steel exhibits an iridescent sheen like that observed on the surface of 'mother-of-pearl'. In both cases, the ridged structure of the surface of the material causes white light to be 'unscrambled' into the original colours of the spectrum.

Let us now recapitulate the main stages in the foregoing process of solidification and cooling of the 0·4% carbon steel.

(1) Solidification is complete at S_1, and the structure consists of uniform austenite.

(2) This austenite begins to transform to ferrite at U_1, the upper critical temperature of this steel (about 825°C).

(3) At 723°C (the lower critical temperature for all steels), formation of primary ferrite ceases, and, as the austenite is now saturated with carbon, the eutectoid pearlite is produced as alternate layers of ferrite and cementite.

(4) Below 723°C, there is no further significant change in the structure.

11.46 A steel which contains exactly 0·83% carbon will begin to solidify as austenite at about 1440°C (Q_2), and be completely solid at approximately 1310°C (S_2). For a steel of this composition, the upper critical and lower critical temperatures coincide at E (723°C); so that no change in the uniformly austenitic structure occurs until a temperature slightly below

723°C is reached, when the austenite will transform to pearlite by precipitating alternate layers of ferrite and cementite. The final structure will be entirely pearlitic (fig. 11.4 (iv)).

11.47 Now let us consider the solidification and cooling of a steel containing, say, 1·2% carbon. This alloy will begin to solidify at approximately 1400°C (Q_3), by depositing dendrites of austenite; and these will grow as the temperature falls, until at about 1230°C (S_3) the structure will be uniform solid austenite. No further change in the structure occurs until the steel reaches its upper critical temperature, at about 900°C (U_3). Then, needle-like crystals of cementite begin to form, mainly at the grain boundaries of the austenite (fig. 11.4 (vi)). In this case, the remaining austenite becomes less rich in carbon, because the carbon-rich compound, cementite, has separated from it. This process continues, until at 723°C the remaining austenite contains only 0·83% carbon (E). This is, of course, the eutectoid composition; so, at a temperature just below 723°C, the remaining austenite transforms to pearlite, as in the previous two cases.

Thus the structure of a carbon steel which has been allowed to cool fairly slowly from any temperature above its upper critical temperature will depend upon the carbon content as follows.

(1) Hypo-eutectoid steels; that is those containing *less than* 0·83% carbon—primary ferrite and pearlite (fig. 11.4 (iii)).

(2) Eutectoid steels, containing *exactly* 0·83% carbon—completely pearlitic (fig. 11.4 (iv)).

(3) Hyper-eutectoid steels; that is those containing *more than* 0·83% carbon—primary cementite and pearlite (fig. 11.4 (vii)).

11.48 Naturally, the proportion of primary ferrite to pearlite in a hypo-eutectoid steel, and also the proportion of primary cementite to pearlite in a hyper-eutectoid steel, will vary with the carbon content, as indicated in fig. 11.6. This diagram summarises the structures, mechanical properties, and uses of plain-carbon steels which have been allowed to cool slowly enough for equilibrium structures to be produced.

The heat-treatment of steel

11.50 The technology of steel is both a tribute to Man's ingenuity, and, at the same time, a manifestation of the wonders of Nature. Depending upon its composition, steel has a wide range of properties such as are found in no other engineering alloy. Steel can be a soft, ductile material suitable for a wide range of forming processes, or it can be the hardest and strongest metallurgical material in use. This enormous range of properties is controlled by varying both the carbon content and the programme of heat-treatment. Structural effects of the type obtained by the heat-treatment of steel would not be possible were it not for the natural phenomenon of allotropy exhibited by the element iron. It is the transformation from a

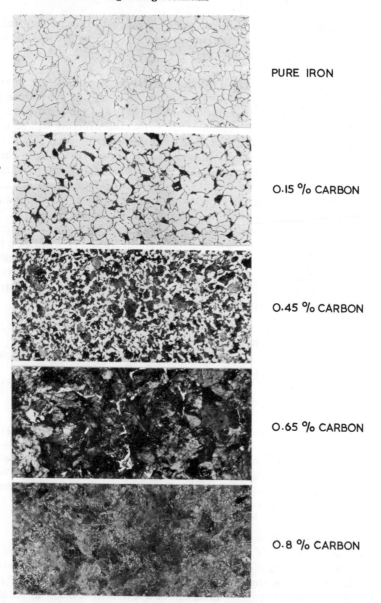

PURE IRON

0.15 % CARBON

0.45 % CARBON

0.65 % CARBON

0.8 % CARBON

Plate 11.1—This series of photomicrographs depicts steels of varying carbon-content, in the normalised condition. As the carbon-content increases, so does the relative proportion of pearlite (dark), until with 0·8% carbon the structure is almost entirely pearlitic. The light areas consist of primary ferrite. The magnification (×100) is not high enough to reveal the laminated nature of the pearlite. In a similar way, craters on the surface of the moon are not visible unless the latter is viewed through a low-power telescope.

face-centred cubic structure to one which is basically body-centred cubic, occurring at 910°C when iron cools, that makes it possible to heat-treat these iron-carbon alloys.

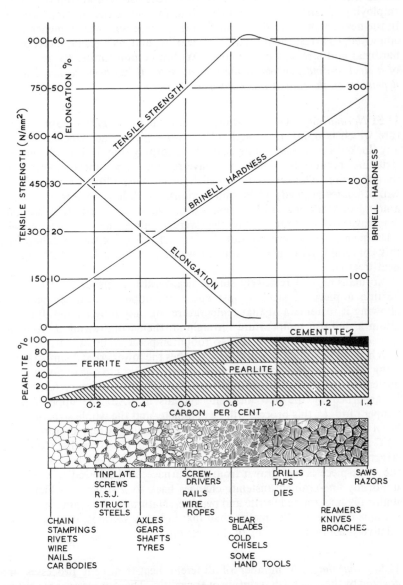

Fig. 11.6—A diagram showing the relationship between carbon content, mechanical properties, microstructure, and uses of plain-carbon steels which have been slowly cooled from above their upper critical temperatures.

There is not just *one* heat-treatment process, but many which can be applied to steels. In the processes we shall deal with in this chapter, the object of the treatment is to obtain a pearlite type of structure; that is, one in which the steel has been allowed to reach structural equilibrium by employing a fairly slow rate of cooling following the heating process. In the next chapter, we shall deal with those processes where quenching is employed, to arrest the formation of pearlite, and, as a result, increase the hardness and strength of the steel. Within certain limits, *the properties of a steel are independent of the rate at which it has been heated, but are dependent on the rate at which it was cooled.*

11.51 *Normalising.* The main purpose in normalising is to obtain a structure which is uniform throughout the work-piece, and which is free of any 'locked-up' stresses. For example, a forging may lack uniformity in structure, because its outer layers have received much more deformation than the core. Thicker sections, which have received little or no working, will be coarse-grained; whilst thin sections, which have undergone a large amount of working, will be fine-grained. Moreover, those thin sections may have cooled so rapidly that they were, in effect, cold-worked, and will consequently be suffering residual stress. If a forging were machined in this condition, its dimensions might well be unstable during subsequent heat-treatment.

Normalising is a relatively simple heat-treatment process. It involves heating a piece of steel to just above its upper critical temperature, allowing it to remain at that temperature for only long enough for it to attain a uniform temperature throughout, then withdrawing it from the furnace, and allowing it to cool to room temperature in still air.

When, on heating, the work-piece reaches the lower critical temperature, L (fig. 11.7), the pearlitic part of the structure changes to one of *fine-grained* austenite; and as the temperature rises, the remaining primary ferrite will be absorbed by the new austenite crystals until, at the upper critical temperature, U, this process will be complete, and the whole structure will be of uniformly fine-grained austenite. In practice, a temperature about 30°C above the upper critical temperature is used, to ensure that the whole structure has reached a temperature just above the upper critical. When the work-piece is withdrawn from the furnace, and allowed to cool, the uniformly fine-grained austenite changes back to a structure which is of uniformly fine-grained ferrite and pearlite. Naturally, the grain size in thin sections may be a little smaller than that in heavy sections, because of the relatively fast rate of cooling of thin sections.

11.52 *Annealing.* A number of different heat-treatment processes are covered by the general description of annealing. These processes are applied to different steels of widely ranging carbon content. The three principal annealing processes will be discussed here.

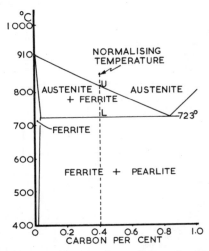

Fig. 11.7—The normalising temperature of a medium-carbon steel in relation to the equilibrium diagram.

11.52.1 *Annealing of castings.* Sand castings in steel commonly contain about 0·3% carbon; so that a structure consisting of ferrite and pearlite is obtained. Such a casting, particularly if massive, will cool very slowly in the sand mould. Consequently, its grain size will be somewhat coarse, and it will suffer from brittleness because of the presence of what is known as a 'Widmanstätten' structure. This consists of a directional plate-like formation of primary ferrite grains along certain crystal planes in the original austenite (fig. 11.8). Since fracture can easily pass along these ferrite plates, the whole structure is rendered brittle as a result.

The annealing process which is applied in order to refine such a structure is fundamentally similar to that described under 'normalising' (11.51); that is, the casting is heated to just above its upper critical temperature, so that the coarse grain structure is replaced by one of fine-grained austenite. On cooling, this gives rise to a structure of fine-grained ferrite and pearlite. It is in the cooling stage where the two processes differ. Whereas air-cooling is employed in normalising; in this process the casting is allowed to cool with the furnace. This ensures the complete removal of all casting stresses, without causing a substantial increase in grain size over that obtained by normalising. Whilst the tensile strength is not greatly improved by this treatment, both toughness and ductility are considerably increased; so the casting becomes more resistant to mechanical shock.

11.52.2 *Spheroidisation annealing.* Although this may appear an onerous title, it is hoped that its meaning will become clear in the following paragraphs. Essentially it is an annealing process which is applied to high-carbon steels in order to improve their machinability, and, in some cases, to make them amenable to cold-drawing.

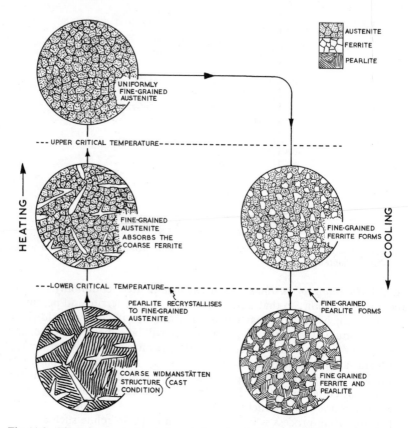

Fig. 11.8—The refinement of grain in a steel casting during a suitable annealing process.

It is an annealing process which is carried out *below* the lower critical temperature of the steel; consequently, no phase change is involved, and we do not need to refer to the equilibrium diagram. The work-piece is held at a temperature between 650 and 700°C for twenty-four hours or more. The pearlite, which of course is still present in the structure at this temperature, undergoes a physical change in pattern—due to a surface-tension* effect at the surface of the *cementite* layers within the pearlite. These layers break up into small plates, due to a tendency of the surface to shrink. This effect causes the plates to become gradually more spherical in form (fig. 11.9)—that is, they spheroidise, or 'ball up'.

When this condition has been reached, the charge is generally allowed to cool in the furnace. Steel is more easily machined in this state, since stresses set up by the pressure of the cutting edge cause minute chip cracks

* In a similar manner, surface tension causes water to form rounded droplets when it is spilled on a very dusty floor.

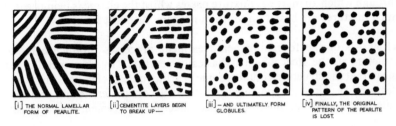

[i] THE NORMAL LAMELLAR FORM OF PEARLITE.   [ii] CEMENTITE LAYERS BEGIN TO BREAK UP —   [iii] — AND ULTIMATELY FORM GLOBULES.   [iv] FINALLY, THE ORIGINAL PATTERN OF THE PEARLITE IS LOST.

Fig. 11.9—The spheroidisation of pearlitic cementite during a sub-critical annealing process.

to form (21.20) in advance of the cutting edge. This is a standard method of improving machinability (21.33).

11.52.3 *Annealing of cold-worked steel.* Like the spheroidisation treatment described above, this also is a sub-critical annealing process. It is employed almost entirely for the softening of cold-worked mild steels, in order that they may receive further cold-work. Such cold-worked materials must be heated to a temperature above the minimum which will cause recrystallisation to take place. Again, the equilibrium diagram is not involved, and the reader should *not* confuse this recrystallisation temperature with the lower critical temperature. The latter is at 723°C; whilst the recrystallisation temperature varies according to the amount of previous cold-work the material has received, but is usually about 550°C. Consequently, stress-relief annealing of mild steel usually involves heating the material at about 650°C for one hour.

This causes the distorted ferrite crystals to recrystallise (fig. 11.10); so that the structure becomes soft again, and its capacity for cold-work is regained. Since the cold-working of mild steel is usually confined to the finishing stages of the product, any annealing is generally carried out in a controlled atmosphere, in order to avoid oxidation of the surface of the charge. The furnace used generally consists of an enclosed 'retort'

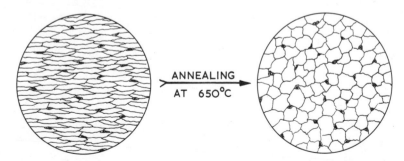

ANNEALING AT 650°C

Fig. 11.10—Annealing of cold-worked mild steel causes recrystallisation of the distorted ferrite; so producing new ferrite crystals, which can again be cold-worked.

through which an inert gas mixture passes whilst the charge is being heated. Such mixtures are based on 'burnt ammonia' or burnt town gas.

11.60 All of the foregoing heat-treatment processes produce a micro-structure in the steel which is basically pearlitic. By this, metallurgists mean that the structure contains some pearlite (unless, of course, the steel is dead-mild). Thus, a hypo-eutectoid steel will contain ferrite and pearlite, a eutectoid steel all pearlite, and a hyper-eutectoid steel cementite and pearlite.

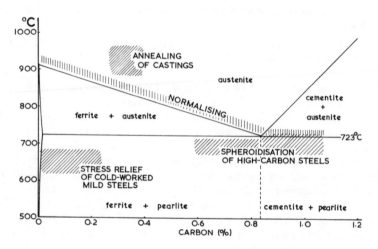

Fig. 11.11—Temperature ranges of various annealing and normalising treatments for carbon steels.

Figure 11.11 summarises the temperature ranges at which these treatments are carried out, and also indicates the carbon-contents of steels most commonly involved in the respective processes.

Chapter Twelve
The Heat-treatment of Plain-carbon Steels

12.10 Some of the simpler heat-treatments applied to steel were described in the previous chapter. In the main, they were processes in which the structure either remained or became basically pearlitic as a result of the treatment. Here we shall deal with those processes which are perhaps better known because of their wider use, namely hardening and tempering.

Almost any schoolboy knows that a piece of carbon steel can be hardened by heating it to redness, and then plunging it into cold water. The Ancients knew this too, and no one can tell us who first hardened steel. Presumably such knowledge came by chance, as indeed did most knowledge in days before systematic research methods were instituted. One can well imagine that sooner or later some prehistoric metal-worker would heat an iron implement among the glowing charcoal of a fire, and then, seeking to cool the implement quickly, would plunge it into water.

12.11 Although the fundamental *technology* of hardening steel has been well established for centuries, the *scientific principles* underlying the process were for long a subject for argument and conjecture. More than a century ago, Professor Henry Sorby of Sheffield began his examinations of the microstructures of steels, but it is only in recent years that convincing explanations of the phenomena of hardening have been forthcoming.

Principles of hardening
12.20 If a piece of steel containing sufficient carbon is heated until its structure is austenitic—that is, until its temperature is above the upper critical temperature—and is then quenched (cooled quickly), it becomes considerably harder than it would be were it cooled slowly.

Generally, when a metallic alloy is quenched, there is a tendency to suppress any changes in structure which might otherwise take place if the alloy were cooled slowly. In other words, it is possible to 'trap' or 'freeze in' a metallic structure which existed at a higher temperature, and so preserve it for examination at room temperature. Metallurgists often use this technique when plotting their equilibrium diagrams, and it is also used industrially, as, for example, in the solution treatment of some aluminium alloys (17.70).

12.21 Clearly, things do not happen in this way when we quench a steel. Austenite, which is the phase present in a steel above its upper critical temperature, is a soft, malleable material—which is why steel is generally shaped by hot-working processes. Yet when we quench austenite, instead of trapping the soft malleable structure, a very hard, brittle structure is produced, which is most unlike austenite. Under the microscope, this structure appears as a mass of uniform needle-shaped crystals, and is known as *martensite*. Even at very high magnifications, no pearlite can be seen; so we must conclude that all of the cementite (which is one of the components of pearlite) is still dissolved in this martensitic structure. So far, this is what we would expect. However, investigations using X-ray methods tell us that, although the rapid cooling has prevented the formation of pearlite, *it has not prevented the allotropic change from face-centred cubic to body-centred cubic*.

Ferrite is fundamentally body-centred cubic iron which normally will dissolve no more than 0·006% carbon at room temperature (see fig. 11.4). Thus the structure of martensite is one which is essentially ferrite very much super-saturated with carbon (assuming that the steel we are dealing with contains about 0·5% carbon). It is easy to imagine that this large amount of carbon remaining in solid solution in the ferrite causes considerable distortion of the internal crystal structure of the latter. Such distortion will tend to prevent slip from taking place in the structure. Consequently, large forces can be applied, and no slip will be produced. In other words, the steel is now hard and strong.

In order to obtain the hard martensitic structure in a steel, it must be cooled quickly enough; that is, at a speed which is at least as fast as *the critical cooling-rate*. If the steel is cooled at a rate slower than this, then the structure will be less hard, because some of the carbon has had the opportunity to precipitate. Under the microscope, some dark patches will be visible among the martensite needles. These are due to the precipitation of some tiny particles of cementite, and the structure so produced is called *bainite*, after Dr E. C. Bain—the American metallurgist who did much of the original research into the relationship between structure and rate of cooling of steels. Bainite is of course softer than martensite, but it is tougher and more ductile. Even slower rates of cooling will give structures of fine pearlite. The critical cooling-rate of a particular steel is therefore the slowest rate at which it can be cooled and a hard martensitic structure obtained.

12.22 In practice, factors such as the composition, size, and shape of the component to be hardened govern the rate at which it shall be cooled. Generally, no attempt is made to harden plain-carbon steels, which contain less than 0·25% carbon, since insufficient carbon is present to give a worth-while increase in hardness. Large masses of steel of heavy section obviously cool more slowly than small components of thin section when quenched; so, whilst the outer skin may be martensitic, the inner core of a large component may contain bainite or even pearlite. More important still, articles of heavy section will be more liable to suffer from quench-cracking. This is due to the fact that the outer skin changes to martensite a fraction of a second before layers just beneath the surface, which are still austenitic. Since sudden *expansion* takes place at the instant when face-centred cubic austenite changes to body-centred cubic martensite, considerable stress will be set up between the skin and the layers beneath it, and, as the skin is now hard and brittle due to martensite formation, cracks may develop in it.

Design also affects the susceptibility of a component to quench-cracking. Sharp variations in cross-section, and the presence of sharp angles, grooves, and notches are all likely to increase the possibility of quench-cracking, by causing uneven rates of cooling throughout the component.

12.23 The rate at which a quenched component cools is governed by the quenching medium and the amount of agitation it receives during quenching. The following media are commonly used, and are arranged in order of quenching speeds.

> 5% caustic soda solution
> 5–20% brine
> cold water
> warm water
> mineral oil
> animal oil
> vegetable oil

The very drastic quenching resulting from the use of caustic soda solution or brine is used only when extreme hardness is required in components of relatively simple shape. For more complex shapes, it would probably be better to use a low-alloy steel, which, as we shall see later, has a much lower critical cooling-rate, and can therefore be hardened by quenching in oil. Mineral oils used for quenching are a by-product of petroleum refining; whilst vegetable oils include those obtained from linseed and cottonseed. Animal oils are obtained from the blubber of seal and whale, though intensive hunting of the latter in recent years has led to its near extinction.

The hardening process
12.30 To harden a hypo-eutectoid steel component, it must be heated to a temperature of 30–50°C above its upper critical temperature, and then quenched in some medium which will produce in it the required rate of cooling. The medium used will depend upon the composition of the steel, the size of the component, and the ultimate properties required in it. Symmetrically shaped components, such as axles, are best quenched 'end-on', and all components should be violently agitated in the medium during the quenching operation.

12.31 The procedure in hardening a hyper-eutectoid steel is slightly different. Here a quenching temperature about 30°C above the *lower critical* temperature is generally used. In a hyper-eutectoid steel, primary cementite is present, and, on cooling from above the upper critical temperature, this

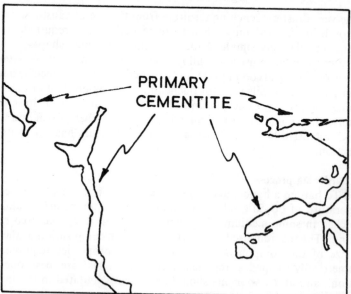

PRIMARY
CEMENTITE

Plate 12.1—A high-carbon tool steel (1·2% C) in the cast condition (×1000).
Since this steel contains more than 0·83% carbon, its structure shows some primary cementite (indicated on the lower 'key' diagram). The remainder of the structure consists of typical lamellar pearlite, comprising layers of cementite sandwiched between layers of ferrite which have tended to join up and so form a continuous background or 'matrix'.

primary cementite tends to precipitate as long, brittle needles along the grain boundaries of the austenite. This type of structure would be very unsatisfactory; so its formation is prevented by continuing to forge the steel whilst this primary cementite is being deposited—that is, between the upper and lower critical temperatures. In this way, the primary cementite is broken down into globules during the final stages of shaping the steel. During the subsequent heat-treatment, it must never be heated much more than 30°C above the lower critical temperature; or there will be a tendency for primary cementite to be absorbed by the austenite, and then precipitated again as long brittle needles on cooling. When a hyper-eutectoid steel has been correctly hardened, its structure should consist of particles of very hard cementite (fig. 12.1) in a matrix of hard martensite.

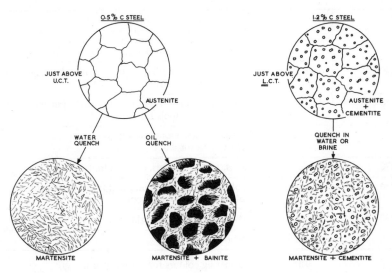

Fig. 12.1—Typical microstructures produced when quenching both medium-carbon and tool steels in their appropriate media.

Tempering
12.40 A fully hardened carbon steel is relatively brittle, and the presence of quenching stresses makes its use in this condition inadvisable unless extreme hardness is required. For these reasons, it is usual to reheat, or 'temper', the quenched component; so that stresses are relieved, and, at the same time, brittleness and extreme hardness are reduced.

As we have seen, the martensitic structure in hardened steel consists essentially of ferrite which is heavily super-saturated with carbon. By heating such a structure to a high enough temperature, we shall enable it to begin to return to equilibrium, by precipitating carbon in the form of tiny particles of cementite.

12.41 On heating the component up to 200°C, no change in the micro-structure occurs, though quenching stresses are relieved to some extent. At about 230°C, tiny particles of cementite are precipitated from the martensite, though these are so small that they are difficult to see with an ordinary microscope. Generally, the microstructure appears somewhat darker, but still retains the shape of the original martensite needles. This type of structure persists as the temperature is increased to about 400°C, with more and more tiny cementite particles being precipitated, and the steel becoming progressively tougher—though softer than the original martensite. The structure so produced was commonly known as *troostite*.

Temperature (°C)	Colour	Types of component
220	pale yellow	scrapers, hack-saws, light turning-tools
230	straw	hammer faces, screwing-dies for brass, planing- and slotting-tools, razor-blades
240	dark straw	shear blades, milling-cutters, drills, boring-cutters, reamers, rock-drills
250	light brown	penknife blades, taps, metal shears, punches, dies, woodworking tools for hardwood
260	purplish-brown	plane blades, stone-cutting tools, punches, reamers, twist-drills for wood
270	purple	axes, augers, gimlets, surgical tools, press-tools
280	deeper purple	cold-chisels (for steel and cast iron), chisels for wood, plane-cutters for softwood
290	bright blue	cold-chisels (for wrought iron), screwdrivers
300	darker blue	wood-saws, springs

Table 12.1—Tempering colours for carbon steels.

12.42 Tempering at temperatures above 400°C causes the cementite particles to coalesce (or fuse together) to such an extent that they can be seen clearly at magnifications of about ×500. At the same time, more cementite is precipitated. The structure, which is relatively granular in appearance, was known as *sorbite*. It must be emphasised that there is no fundamental difference between troostite and sorbite, since both are formed by precipitation of cementite from martensite; and there is no definite temperature where troostite formation ceases and formation of sorbite begins. Naturally, sorbite is softer and tougher than troostite, because still more carbon has been precipitated from the original martensite structure.

The names 'troostite' and 'sorbite' are now obsolete, and modern metallurgists prefer to describe such structures as 'tempered martensite', mentioning the temperature used during the tempering process.

Generally speaking, low temperatures (200–300°C) are used for temper-ing various types of high-carbon steel tools where hardness is the prime consideration, higher temperatures (400–600°C) being used for tempering stress-bearing medium-carbon constructional steels where strength, tough-ness, and general reliability are more important.

12.43 Furnaces used for tempering are usually of the batch type, in which the charge is carried in a wire basket through which hot air circulates. By this method, the necessarily accurate temperature can easily be main-tained. The traditional method of treating tools is to 'temper by colour', and this still provides an accurate and reliable method of dealing with plain-carbon steels. After the tool has been quenched, its surface is first cleaned to expose bright metal. The tool is then slowly heated until the thin oxide skin which forms on the surface attains the correct colour (table 12.1). It should be noted that this technique applies only to plain-carbon steels, since some of the alloy steels, particularly those containing chromium, do not oxidise readily. A summary of typical heat-treatment programmes, and uses of the complete range of plain-carbon steels, is given in table 12.2.

Mass effect and hardenability
12.50 In order to harden a piece of steel completely, it must be cooled quickly from its austenitic state. The rate of cooling must be as great as, or greater than, the 'critical rate'; otherwise the section will not be completely martensitic. Clearly, this will not be possible for a work-piece of very heavy section; for, whilst the outer skin will cool at a speed greater than the critical rate, the core will not. Consequently, whilst the outer shell may be of hard martensite, the core may be of bainite, or even fine pearlite.

12.51 This phenomenon is generally referred to as the 'mass effect' of heat-treatment, and plain-carbon steel is said to have a 'shallow depth of hardening', or, alternatively, 'a poor hardenability'. Whilst a rod of, say, 12 mm diameter in plain-carbon steel can be water-quenched successfully, and have a martensitic structure throughout; one of, say, 30 mm diameter will have a core consisting of bainite, and only the outer shell will be martensitic. The influence of sectional thickness and quenching medium used on the structure produced in a low-alloy steel is illustrated in fig. 12.2.

12.52 For some applications, this variation in structure across the section will not matter, since the outer skin will be very hard, and the core reasonably tough; but in cases where a component is to be quenched and then tempered for use as a stress-bearing member, it is essential that the structure obtained by quenching is uniform throughout.

Type of steel	Carbon %	Heat-treatment		
		Hardening temp. (°C)	Quenching medium	Tempering temp. (°C)
dead-mild	Up to 0·15	These steels do not respond to heat-treatment, because of their low carbon-content		
mild	0·15 to 0·25			
medium-carbon	0·25 to 0·35	880 to 850*	Oil or water, depending upon type of work	Temper as required
	0·35 to 0·45	870 to 830*		
	0·45 to 0·60	850 to 800*	Oil, water, or brine, depending upon type of tool	
high-carbon tool	0·60 to 0·75	820 to 800*	Water or brine. Tools should not be allowed to cool below 100°C before tempering	275–300
	0·75 to 0·90	800 to 820		240–250
	0·90 to 1·05	780 to 800		230–250
	1·05 to 1·20	760 to 780		230–250
	1·20 to 1·35	760 to 780		240–250
	1·35 to 1·50	760 to 780		200–230

* The *higher* temperature for the *lower* carbon-content.

Table 12.2—Heat-treatments and typical uses of plain-carbon steels.

Ruling section

12.60 Fortunately, the addition of alloying elements to a steel reduces its critical rate so that such a steel can be oil-quenched in thin sections; or, alternatively, much heavier sections can be water-quenched. Thus an alloy steel generally has a much greater depth of hardening than has a plain-carbon steel of similar carbon content.

However, it would be a mistake to assume that *any* alloy steel in *any* thickness will harden right through when oil-quenched from above its upper critical temperature. The low-alloy steel used as an example in

Typical uses

nails, chains, rivets, motor-car bodies

structural steels (RSJ), screws, tinplate, drop-forgings, stampings, shafting, free-cutting steels

couplings, crankshafts, washers, steering arms, lugs, weldless steel tubes

crankshafts, rotor shafts, crank pins, axles, gears, forgings of many types

hand-tools, pliers, screwdrivers, gears, die-blocks, rails, laminated springs, wire ropes

hammers, dies, chisels, miners' tools, boilermakers' tools, set-screws

cold-chisels, blacksmiths' tools, cold-shear blades, heavy screwing dies, mining drills

hot-shear blades, taps, reamers, threading and trimming dies, mill-picks

taps, reamers, drills, punches, blanking-tools, large turning-tools

lathe tools, small cold-chisels, cutters, drills, pincers, shear blades

razors, wood-cutting tools, surgical instruments, drills, slotting-tools, small taps

fig. 12.2 illustrates this point. This steel has a critical cooling rate just a little lower than that of a plain-carbon steel; but, as the diagram shows, sections much over 14 mm in diameter will not harden completely unless quenched in brine, and the maximum diameter which can be hardened completely—even with this drastic treatment—is only about 30 mm.

12.61 In order to prevent the misuse of steels by those who imagine that an alloy steel can be hardened to almost any depth, both the British Standards Institution and manufacturers now specify limiting ruling sections for

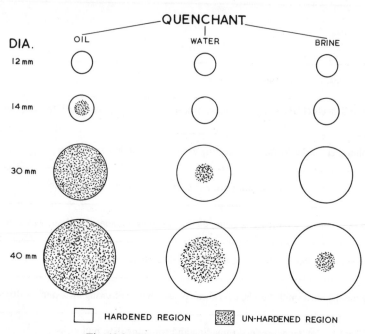

Fig. 12.2—The mass effect of heat-treatment.
Specimens of different diameter have been quenched in different media, and the depth of hardening assessed in each case. A low-alloy steel containing 0·25% C, 0·6% Mn, 0·2% Ni, and 0·2% Mo was used.

each particular composition of steel. The limiting ruling section is quoted as the *maximum diameter* which can be heat-treated (under conditions of quenching and tempering suggested by the manufacturer) for the stated mechanical properties to be obtained. As an example, the following table shows a set of ruling sections for a low alloy steel (BS 970/530M40), along with a manufacturer's suggested heat-treatments, and the corresponding BS specifications in respect of tensile strength.

Limiting ruling section (mm)	Suggested heat-treatment	Tensile strength (N/mm²)
100	oil harden	695
62·5	from 850°C,	770
29	and temper	850
	at 650°C	

In BS 970, formulae are given for deriving the equivalent diameters of rectangular and other bars, so that information as to the ruling section of the material can be correctly applied.

The Jominy end-quench test
12.70 This test is of considerable value in assessing the hardenability of a

steel. A standard test-piece (fig. 12.3 (ii)) is heated to above the upper critical temperature of the steel; that is, until it becomes completely austenitic. It is then very quickly transferred from the furnace, and immediately dropped into position in the frame of the apparatus shown in

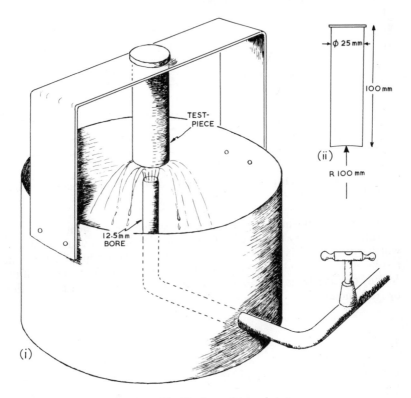

Fig. 12.3—The Jominy end-quench test.
(i) A simple type of apparatus in which to conduct the test. The end of the water pipe is 12 mm below the bottom of the test-piece, but the 'free height' of the water jet is 63 mm.
(ii) A typical test-piece.

fig. 12.3 (i). Here it is quenched at one end only, by a standard jet of water at 25°C; thus, different rates of cooling are obtained along the length of the test-piece. When the test-piece has cooled, a 'flat' approximately 0·4 mm deep is ground along the length of the bar; and hardness determinations are made every millimetre along the length, from the quenched end. The results are then plotted (fig. 12.4).

12.71 These curves show that a low nickel–chromium steel hardens to a greater depth than a plain-carbon steel of similar carbon content; whilst a

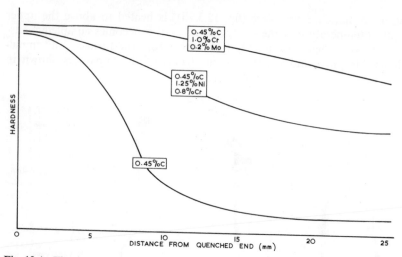

Fig. 12.4—The depth of hardening of various steels of similar carbon content, as shown by the Jominy test.

chromium–molybdenum steel hardens to an even greater depth.

Whilst the Jominy test gives a good indication as to how deeply a steel will harden, there is no simple mathematical relationship between the results of this test and the ruling section of a steel. Using the results of the Jominy test as a basis, it is often more satisfactory to find the ruling section by trial and error.

Chapter Thirteen
Alloy Steels

13.10 The first deliberate attempt to develop an alloy steel was made by Sir Robert Hadfield, less than a century ago; and most of these materials are products of the present century. Possibly the greatest advance in this field was made in 1900, when Taylor and White, two engineers with the Bethlehem Steel Company in the USA, introduced the first high-speed steel, and so helped to increase the momentum of the Industrial Revolution and, in particular, what has since been known as 'mass-production'.

So-called plain-carbon steels contain up to $1 \cdot 0\%$ manganese, which is the residue of that added to deoxidise and desulphurise the steel just before casting. Consequently, a steel is not classed as an alloy steel unless it contains more than $1 \cdot 0\%$ manganese, or deliberate additions of other elements.

Although there are a number of alloy steels with special properties, such as the stainless steels, and heat-resisting steels, the main purpose of alloying is to improve the existing properties of carbon steels, making them more adaptable and easier to heat-treat successfully. In fact, one of the most important and useful effects of alloying was mentioned in the previous chapter (12.60)—the improvement in 'hardenability'. Thus an alloy steel can be successfully hardened by quenching in oil, or even in an air blast, with less risk of distortion or cracking of the component than is associated with water-quenching.

13.11 Alloying elements (or 'alloys' as they are often called) can be divided into two main groups.

(1) Those which strengthen and toughen the steel by dissolving in the ferrite. These elements are used mainly in constructional steels, and include nickel, manganese, small amounts of chromium, and even smaller amounts of molybdenum.

(2) Alloying elements which combine chemically with some of the carbon in the steel, to form carbides which are much harder than iron carbide (cementite). These elements are used mainly in tool steels, die steels, and the like. They include chromium, tungsten, molybdenum, and vanadium.

13.12 Other alloying elements which are added in small amounts and for special purposes include titanium, niobium, aluminium, copper, and silicon. Even sulphur, normally regarded as the steel-maker's greatest enemy, is utilised in free-cutting 'bright-drawn' steels (21.21 and 21.40).

Alloy steels may be classified into three main groups:

(1) constructional steels, which are generally used for machine parts highly stressed in tension or compression;

(2) tool steels, requiring great hardness and, in some cases, resistance to softening by heat;

(3) special steels; for example, stainless steels and heat-resisting steels.

Constructional steels

13.20 Whilst the 'nickel-chrome' steels are the best known in this group, other alloy steels containing the elements nickel and chromium singly are also important.

13.21 *Nickel steels.* Nickel increases the strength of a steel by dissolving in the ferrite. Its main effect, however, is to increase toughness by limiting grain-growth during heat-treatment processes. For this reason, up to $5 \cdot 0\%$

BS 970 : Part 2 : 1970 specification	Composition (%)	Typical mechanical properties		
		Yield point (N/mm²)	Tensile strength (N/mm²)	Izod (J)
503M40	0·4 C 1·0 Mn 1·0 Ni	500	700	96
—	0·12 C 0·45 Mn 3·0 Ni	510	775	86
—	0·12 C 0·4 Mn 5·0 Ni	600	850	71

Table 13.1—Nickel steels.

BS 970 : Part 2 : 1970 specification	Composition (%)	Typical mechanical properties		
		Yield point (N/mm²)	Tensile strength (N/mm²)	Hardness (brinell)
530M40	0·45 C 0·9 Mn 1·0 Cr	880	990	—
535A99	1·0 C 0·45 Mn 1·4 Cr	—	—	850

Table 13.2—Chromium steels.

nickel is present in some of the better quality steels used for case-hardening.

Unfortunately, nickel does not combine chemically with carbon, and, worse still, tends to make iron carbide (cementite) decompose and so release free graphite. Consequently, nickel steels are always low-carbon steels, or, alternatively, medium-carbon steels with very small amounts of nickel.

13.22 *Chromium steels.* When chromium is added to a steel, some of it dissolves in the ferrite (which is strengthened), but the remainder forms chromium carbide. Since chromium carbide is harder than ordinary iron carbide (cementite), the hardness of the steel is increased. Because chromium forms stable carbides, these steels may contain $1\cdot0\%$ or even more of carbon.

The main disadvantage of chromium as an alloying element is that, unlike

Heat-treatment	Uses
Oil-quench from 850°C, temper between 550°C and 650°C	Crankshafts, axles, other parts in the motor-car industry and in general engineering
After carburising: refine grain by an oil-quench from 860°C, then harden by a water-quench from 770°C	*A case-hardening steel:* crown-wheels, differential pinions, cam-shafts
After carburisation: refine grain by an oil-quench from 850°C, then harden by an oil-quench from 760°C	*Heavy-duty case-hardened parts:* bevel-pinions, gudgeon-pins, gear-box gears, worm-shafts

Heat-treatment	Uses
Oil-quench from 860°C, temper at 550–700°C	Agricultural machine parts, machine-tool components, parts for concrete and tar mixers, excavator teeth, automobile axles, connecting rods and steering arms, spanners
Oil-quench from 810°C, temper at 150°C	Ball- and roller-bearings, roller- and ball-races, cams, small rolls

nickel, it increases grain growth during heat-treatment. Thus, unless care is taken to limit both the temperature and the time of such treatment, brittleness may arise from the coarse grain produced.

As indicated by the uses mentioned in table 13.2, these low-chromium steels are important because of their increased hardness and wear-resistance.

BS 970: *Part* 2: 1970 *specification*	*Composition* (%)	*Typical mechanical properties*		
		Yield point (N/mm²)	*Tensile strength* (N/mm²)	*Izod* (J)
653M31	0·3 C 0·6 Mn 3·0 Ni 1·0 Cr	820	930	68
—	0·3 C 0·5 Mn 4·25 Ni 1·25 Cr	990 820	1020 980	13* 73
—	0·15 C 0·4 Mn 3·5 Ni 0·8 Cr	—	—	—

* Low impact value, due to temper brittleness caused by tempering the steel at 500°C
Table 13.3—Nickel–chromium steels.

BS 970: *Part* 2: 1970 *specification*	*Composition* (%)	*Typical mechanical properties*		
		Yield point (N/mm²)	*Tensile strength* (N/mm²)	*Izod* (J)
817M40	0·4C 0·55 Mn 1·5 Ni 1·1 Cr 0·3 Mo	990	1010	72
830M31	0·3 C 0·65 Mn 3·0 Ni 1·2 Cr 0·35 Mo	880	980	108
835M30	0·3 C 0·5 Mn 4·25 Ni 1·25 Cr 0·3 Mo	1450	1700	37

Table 13.4—Nickel–chromium–molybdenum steels.

13.23 *Nickel–chromium steels.* In the foregoing sections dealing with nickel and chromium, it was seen that in some respects the two metals have opposite effects on the properties of a steel. Thus, whilst nickel is a grain-refiner, chromium tends to cause grain-growth. On the other hand, whilst chromium is a carbide-stabiliser, nickel tends to cause carbides to break

Heat-treatment	Uses
Oil-quench from 830°C, temper at 550–650°C and cool in oil to avoid temper brittleness.	Highly-stressed parts in aero- and automobile-engineering; differential shafts, stub-axles and connecting rods
Air-harden from 820°C, temper at 200°C for max. hardness or at 650°C for max. ductility. Do *not* temper between 250°C and 580°C as this causes temper brittleness.	Parts of complex shape where oil-quenching might cause distortion, e.g. gears and driving shafts
—	*Case-hardening:* parts requiring a glass-hard surface and toughness of core: high-duty gears, worm-gears, crown-wheels, clutch gears

Heat-treatment	Uses
Oil-quench from 840°C, temper at 600°C	Differential shafts, crank-shafts and other highly-stressed parts. (If tempered at 200°C, it is suitable for machine-tool and automobile gears)
Oil-quench from 830°C, temper at 600°C	Thin sections where maximum shock-resistance and ductility are required, e.g. connecting rods, inlet-valves, cylinder-studs, valve-rockers
Air-harden from 830°C, temper at 150–200°C	An air-hardening steel for aero-engine connecting rods, valve-mechanisms, gears, differential shafts, etc. Suitable for other highly-stressed parts

down, releasing graphite. Fortunately, the beneficial effects of one metal are stronger than the adverse effects of the other, and so it is advantageous to add these metals together to a steel. Generally about two parts of nickel to one of chromium is found to be the best proportion.

In other respects, the two metals, as it were, work together; and so the hardenability is increased to the extent that, with 4·25% nickel and 1·25% chromium, an *air-hardening* steel is produced; that is, one which can be 'quenched' in an air blast, thus making cracking or distortion even less likely than if the steel were oil-quenched. However, for air-hardening, a ruling section of 62·5 mm diameter must be observed, and for greater diameters than this the steel must be oil-quenched if the stated properties are to be obtained.

Unfortunately, these straight nickel-chromium steels suffer from a defect known as 'temper brittleness'. This will be described in the next section.

13.24 *Nickel–chromium–molybdenum steels.* As mentioned above, a severe drawback in the use of straight nickel-chromium steels is that they suffer from a defect known as 'temper brittleness'. This is shown by a serious decrease in toughness (as indicated by a low Izod or Charpy impact value) when a quenched steel is subsequently tempered in the range 250–580°C. Further, if such a steel is tempered at 650°C, it must be cooled quickly through the 'dangerous range' by quenching it in oil, following the tempering process. Although incorrect heat-treatment may lead to this disastrous reduction in impact toughness, the tensile strength and percentage elongation may not be seriously affected. Consequently, a tensile test alone would not reveal the shortcomings of such a steel, and the importance of impact testing in cases like this is obvious.

BS specification	Composition (%)	Typical mechanical properties			
		Yield point (N/mm²)	Tensile strength (N/mm²)	Izod (J)	Brinell
970: *Part* 1: 1970 —150M36	0·35 C 1·5 Mn	510	710	71	—
970: *Part* 2:1970 —608M38	0·35 C 1·6 Mn 0·45 Mo	1000	1130	72	—
— (Hadfield steel)	1·2 C 12·5 Mn	—	—	—	Case-550 Core-200

Table 13.5—Manganese steels.

Fortunately, temper brittleness can be largely eliminated by adding about 0·3% molybdenum to the steel, thus establishing the well known range of 'nickel–chrome–moly' steels.

13.25 *Manganese steels.* Most steels contain some manganese remaining from the deoxidation and desulphurisation processes, but it is only when the manganese content exceeds 1·0% that it is regarded as an alloying element. Manganese increases the strength and toughness of a steel, but less effectively than does nickel. Like all elements, it increases the depth of hardening. Consequently, low-manganese steels are used as substitutes for other, more expensive, low-alloy steels.

Manganese is a metal with a structure somewhat similar to that of austenite at ordinary temperatures; therefore, when added to a steel in sufficient quantities, it tends to stabilise the FCC (austenitic) structure of iron at lower temperatures than is normal for austenite. In fact, if 12·0% manganese is added to a steel containing 1·0% carbon, the structure remains austenitic even after the steel has been *slowly cooled* to room temperature. The curious—and useful—fact about this steel is that, if the surface suffers any sort of mechanical disturbance, its structure immediately changes from soft austenite to extremely hard martensite; hence this high-manganese steel, having a tough austenitic core, and developing a hard martensitic shell, is useful in conditions where both mechanical shock and severe abrasion prevail, as in dredging, earth-moving, and rock-crushing equipment. A further point of interest regarding this steel is that it was one of the very first alloy steels to be developed—by Sir Robert Hadfield in 1882—though little use was made of it until the early days of this century.

Heat-treatment	Uses
Oil-quench from 850°C, temper at 600°C	Automobile and general engineering, as a cheaper substitute for the more expensive nickel-chromium steels
Oil-quench from 850°C, temper at 600°C	Automobile and general engineering, as a cheaper substitute for the more expensive nickel-chromium steels
Finish by quenching from 1050°C to keep carbides in solution—quenching does *not* harden the steel, however	Rock-crushing equipment, buckets, heel-plates and bucket-lips for dredging equipment, earth-moving equipment, trackway crossings and points

BS spec'n	Composition (%)	Hardness (VPN)	Heat-treatment	Uses
526 M60†	0·6 C 0·65 Mn 0·65 Cr	700	Oil-quench from 830°C. Temper: (i) for cold-working tools at 200–300°C (ii) for hot-working tools at 400–600°C.	Blacksmiths' and boiler-makers' tools and chisels, swages, builders', masons' and miners' tools, chuck- and vice-jaws, hot-stamping and forging dies
BD 3*	2·1 C 12·5 Cr	850	Heat slowly to 800°C, then raise to 980°C, and oil-quench. Temper between 150° and 400°C for 30–60 minutes.	Blanking punches, dies, shear blades for hard thin materials, dies for moulding ceramics and other abrasive powders, master-gauges, thread-rolling dies
BH 11*	0·35 C 5·0 Cr 1·25 Mo 0·3 V 1·0 Si	Tempered at 550°C: 600 Tempered at 650°C: 375	Pre-heat to 850°C, then heat to 1000°C. Soak for 10–30 minutes, and air-harden. Temper at 550–650°C for 2 hours.	Hot-forging dies for steel and copper alloys where excessive temperatures are not encountered, extrusion dies for aluminium alloys, pressure and gravity dies for casting aluminium
BD 2A*	1·6 C 13·0 Cr 0·8 Mo 0·5 V	Tempered at 200°C: 800 Tempered at 400°C: 700	Pre-heat to 850°C and then heat to 1000°C. Soak for 15–45 minutes, and quench in oil or air. Temper at 200–400°C for 30–60 minutes	Fine press-tools, deep-drawing and forming dies for sheet metal, wire-drawing dies, blanking dies, punches and shear blades for hard metals
BH 12*	0·35 C 1·0 Si 5·0 Cr 1·5 Mo 0·45 V 1·35 W	—	Pre-heat to 800°C. Soak, and then heat quickly to 1020°C. Air-quench, and then temper for 90 minutes at 540–620°C.	Extrusion dies for aluminium and copper alloys, hot-forming, piercing-, and heading-tools, brass-forging dies
BH 21*	0·3 C 2·85 Cr 0·35 V 10·0 W	—	Pre-heat to 850°C, and then heat rapidly to 1200°C. Oil-quench (or air-quench thin sections). Temper at 600–700°C for 2–3 hours.	Hot-forging dies and punches for making bolts, rivets, etc. where tools reach high temperatures, hot-forging dies, extrusion dies and die-casting dies for copper alloys, pressure die-casting dies for aluminium alloys

Table 13.6—Tool steels and die steels (other than high-speed steels).

† BS 970 : Part 2: 1970 designation, * BS 4659 designation

Tool steels and die steels

13.30 The primary requirement of a tool or die steel is that it shall have considerable hardness and wear-resistance, combined with reasonable mechanical strength and toughness. A plain high-carbon tool steel possesses these properties, but unfortunately its cutting edge softens easily on becoming over-heated during a high-speed cutting process. Similarly, dies which are to be used for hot-forging or extrusion operations cannot be made from plain-carbon steel, which, in the heat-treated state, begins to soften at about 220°C. Consequently, tool steels which work at high speeds, or die steels which work at high temperatures, are generally alloy steels containing one or more of those elements which form very hard carbides—chromium, tungsten, molybdenum, or vanadium. Of these elements, tungsten and molybdenum also cause the steel, once hardened, to develop a resistance to tempering influences, whether from contact with a hot work-piece, or from frictional heat. Thus, either tungsten or molybdenum is present in all high-speed steels, and in most high-temperature die steels.

13.31 *Die steels.* As mentioned above, these materials will contain at least one of the four metals which form hard carbides; whilst hot-working dies will in any case contain either tungsten or molybdenum, to provide the necessary strength and hardness at high temperatures.

The heat-treatment of these steels resembles that for high-speed steels, which will be described in the next section.

13.32 *High-speed steel.* High-speed steel, as we know it, was first shown to an amazed public at the Paris Exposition of 1900. A tool was exhibited cutting at a speed of some 0·3 m/s, with its tip heated to redness. Soon after this, it was found that the maximum cutting efficiency was attained with a composition of 18% tungsten, 4% chromium, 1% vanadium, and 0·75% carbon; and this remains possibly the best known general-purpose high-speed steel to this day.

13.33 Since high-speed steel is a complex alloy, containing at least five different elements, it cannot be represented by an ordinary equilibrium diagram (9.70); however, by grouping all the alloying elements together, under the title 'complex carbides', a simplified two-dimensional diagram (fig. 13.1) can be used to explain the heat-treatment of this material.

It will be seen that this diagram still resembles the ordinary iron-carbon diagram in general shape. The main difference is that the lower critical temperature has been raised (alloying elements usually raise or lower this temperature), and the eutectoid point E is now at only 0·25% carbon (instead of 0·83%). All alloying elements cause a shift of the eutectoid point to the left, for which reason alloy steels generally contain less carbon than the equivalent plain-carbon steels.

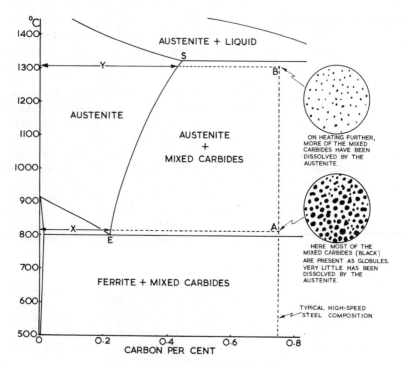

Fig. 13.1—A modified equilibrium diagram for high-speed steel.

If the 'typical high-speed steel composition' is heated to '*A*', only an amount of mixed carbides equivalent to '*X*' is dissolved by the austenite. Hence, on quenching, the steel would be soft, and would *not* resist tempering influences. The maximum amount of mixed carbides which can be safely dissolved (without beginning to melt the tool) is shown by '*Y*', and this involves heating the tool to 1300°C—just short of its melting-point.

13.34 In the normalised condition, a typical high-speed steel contains massive globules of carbide in a matrix (or background) of ferrite. If this is now heated to just above the lower critical temperature (*A* in fig. 13.1) the ferrite will change to austenite, and begin to dissolve the carbide globules. If the steel were quenched from this point, in the manner of a plain-carbon tool steel, the resultant structure would lack hardness, since only 0·25 % carbon, or thereabouts, would be dissolved in the martensite so produced. Moreover, it would not resist tempering influences, as little tungsten would be dissolved in the martensite either. It is therefore necessary to ensure that the maximum amount of tungsten carbide is dissolved in the austenite before the steel is quenched.

The slope of the boundary line *ES* shows that, as the temperature rises, the amount of carbides dissolved in the austenite increases to a maximum at *S*, where the steel begins to melt (approximately 1320°C). Hence, to make sure that the maximum amount of carbide is dissolved before the steel is quenched, a high quenching-temperature in the region of 1300°C is necessary. Since this is just short of the temperature at which melting begins, grain growth will proceed rather quickly. For this reason, a special heat-treatment furnace must be used. This consists of a lower chamber, usually heated by gas, and running at the quenching-temperature of 1300°C; and above this is a preheater chamber, maintained at about 850°C by the exhaust gases which have already circulated around the high-temperature chamber. The tool is first preheated to 850°C, and transferred to the high-temperature compartment, where it will reach the quenching-temperature in a few minutes. In this way, the time of contact between tool and high-temperature conditions is reduced below that which would be necessary were the tool not preheated. At such high temperatures, decarburisation of the tool surface would be serious, so a controlled non-oxidising atmosphere is generally used in the furnace chamber.

BS 4659 spec'n	Composition (%)	Heat-treatment		Hardness (VPN)	Uses
		Quench in oil or air from	Secondary-hardening treatment		
BT 21	0·65 C 4·0 Cr 14·0 W 0·5 V	1300°C	Double temper at 565°C for 1 hour. (The double-temper treatment gives extra hardness, as more austenite transforms to martensite.)	860	General shop practice—all-round work, moderate duties, also for blanking-tools and shear blades.
BT 1	0·8 C 4·5 Cr 18·0 W 1·2 V	1310°C		890	Lathe-, planer-, and shaping-tools, millers and gear-cutters, reamers broaches, taps, dies, drills, hacksaws, roller-bearings for gas turbines
BT 6	0·8 C 4·25 Cr 20·0 W 1·5 V 0·5 Mo 12·0 Co	1320°C		950	Lathe-, planer-, and shaper-tools, milling-cutters, drills for very hard materials. So-called 'super high-speed' steel, which has maximum hardness and toughness.
BM 1	0·8 C 3·75 Cr 1·6 W 1·25 W 9·0 Mo	1230°C		900	A general-purpose molybdenum-type high-speed steel for drills, taps, reamers, cutters. Susceptible to decarburisation during heat-treatment, which therefore requires careful control.

Table 13.7—High-speed steels.

13.35 As soon as it has reached the quenching-temperature, the tool is quenched in oil or in an air blast (depending upon its size and composition). The resultant structure contains some martensite, but also some soft austenite, because the high alloy-content considerably reduces the rate of transformation. Hence the steel is heated to about 550°C, to promote transformation of this austenite to martensite. This process is known as *secondary hardening*, and gives an increase in hardness from about 700 to over 800 VPN. 'Super high-speed' steels contain up to 12% cobalt, and are harder than the ordinary tungsten types.

13.36 Since molybdenum is now cheaper than tungsten, many modern high-speed steels contain large amounts of molybdenum to replace much of the tungsten. These molybdenum-type steels are reputed to be more difficult to heat-treat successfully, and, whilst they are widely used in the USA, they are less popular in Britain.

Sintered tool materials
13.40 Although these materials are not steels, it is convenient to mention them here, because their properties are similar to those of a high-speed steel, containing as they do particles of a hard constituent in a hard, tough matrix, which will not soften easily.

13.41 *Sintered carbides.* These are based mainly on tungsten carbide and cobalt, and are powder-metallurgy products, the manufacture of which is described elsewhere (7.43). The structure consists of tungsten carbide particles in a tough cobalt matrix. Since these sintered carbides are relatively brittle, they are generally brazed to a steel shank for service.

13.42 *Sintered oxides.* Aluminium oxide, or alumina, is one of the hardest metallic oxides available, and has for many years been used as a polishing powder in metallography. If crystallised so that it forms a solid mass, or, alternatively, if bonded with a suitable medium, alumina can be used as a cutting-tool material. These tool tips contain at least 85% alumina, which is bonded with a form of glass, though other bonding-materials are sometimes used.

Not only do these materials retain hardness and compressive strength at high temperatures, but they also have low-friction properties, and high resistance to both abrasion and chemical attack; consequently, they are often used for cutting at higher speeds than is possible with 'hard metal' tools. They are used principally for cutting tough materials, including plastics, rubber, wood, aluminium, and ceramics.

The tool-piece is attached either by metallising it and then brazing it to the shank, or by bonding it with an epoxy resin adhesive (table 26.1). Mechanical clamping is also used, and these tips are used on the 'throw away' principle, being discarded when they become worn.

Stainless steels

13.50 Although Michael Faraday had attempted to produce stainless steel in as long ago as 1822, it was not until 1912 that Brearley discovered the rust-resisting properties of high-chromium steel.

Chromium imparts the 'stainless' properties to these steels by coating the surface with a thin but extremely dense film of chromium oxide, which effectively protects the surface from further attack. Ordinary steel, on the other hand, becomes coated with a loose, porous layer of rust, through which the atmosphere can pass and cause further corrosion. For this reason, ordinary steel rusts quickly, the top flakes of rust being pushed off by new layers forming beneath.

BS spec'n	Composition (%)	Heat-treatment	Uses
403S17*	0·04 C 0·45 Mn 14·0 Cr	Non-hardening, except by cold-work	'Stainless iron'—domestic articles such as forks and spoons. Can be pressed, drawn, and spun.
420S45*	0·3 C 0·5 Mn 13·0 Cr	Oil- or water-quench (or air-cool) from 960°C. Temper (for cutting) 150–180°C, temper (for springs) 400–450°C.	Specially for cutlery and sharp-edged tools, springs, circlips
302S25*	0·1 C 0·8 Mn 8·5 Ni 18·0 Cr	Non-hardening except by cold-work. (Cool quickly from 1050°C, to keep carbides dissolved.)	Particularly suitable for domestic and decorative purposes
347S17*	0·05 C 0·8 Mn 10·0 Ni 18·0 Cr 1·0 Nb		Weld-decay proofed by the presence of Nb. Used in welded plant where corrosive conditions are severe, e.g. nitric-acid plant

Table 13.8—Stainless steels (*BS 970: Part 4: 1970 designation).

13.51 Much corrosion in metals is of the 'electrolytic' type (22.30). Readers will be familiar with the working of a simple cell, in which a copper plate and a zinc plate are immersed in dilute sulphuric acid (called the 'electrolyte'). As soon as the plates are connected, a current flows, and the zinc plate dissolves ('corrodes') rapidly. In many alloys containing crystals of two different compositions, corrosion of one type of crystal will occur in this electrolytic manner when the surface of the alloy is coated with an electrolyte—which, incidentally, may be rain water. In stainless steels, however, the structure is a uniform solid solution. Since all of the crystals within a piece of the alloy are of the same composition, electrolytic action cannot take place. There are two main types of stainless steel.

(1) The straight chromium alloys, which contain 13% or more of chromium. These steels, provided they contain sufficient carbon, can be heat-treated to give a hard martensitic structure. Stainless cutlery

steel is of this type. Some of these steels, however, contain little or no carbon, and are pressed and deep-drawn to produce such articles as domestic kitchen-sinks, refrigerator parts, beer-barrels, and table-ware.

(2) The '18/8' chromium/nickel steels, which are austenitic even after being cooled slowly to room temperature. This type of steel cannot be hardened (except of course by cold-work), and is used solely for constructional and ornamental work. Much of it is used in chemical plant, where acid-resisting properties are required.

13.52 Although these austenitic stainless steels cannot be hardened by heat-treatment, they are usually 'finished' by quenching from 1050°C. The purpose of this treatment is to prevent the precipitation of particles of chromium carbide, which would occur if the steel were allowed to cool slowly to room temperature.

The precipitation of chromium carbide particles would draw out chromium from the surrounding structure, leaving it almost free of chromium (fig. 13.2) so that rusting would occur in that region. Such corrosion would be due to a combination of electrolytic action and direct attack.

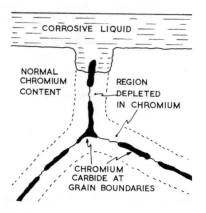

Fig. 13.2—The effect of carbide precipitation on the resistance to corrosion.

Because of the risk of precipitation of chromium carbide, these steels are unsuitable for welding, and suffer from a defect known as 'weld-decay'.

13.53 During welding, some regions of the metal near to the weld will be maintained between 650 and 800°C long enough for chromium carbide to precipitate there (fig. 13.3). Subsequently, corrosion will occur in this area near to the weld. The fault may largely be overcome by adding about 1% of either titanium or niobium. These metals have a great affinity for carbon, which therefore combines with them in preference to chromium. Thus chromium is not drawn out of the structure, which, as a result, remains uniform.

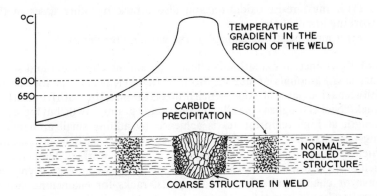

Fig. 13.3—Microstructural changes during welding which lead to subsequent corrosion ('weld decay') in some stainless steels.

Heat-resisting steels

13.60 The main requirements of a steel to be used at high temperatures are:

Composition (%)	Heat-treatment	Maximum working temperature (°C)	Uses
0·4 C 0·2 Si 1·4 Mn 10·0 Cr 36·0 Ni	—	600	Steam-turbine blades and other fittings
0·1 C 0·7 Mn 12·0 Cr 2·5 Ni 1·8 Mo 0·35 V	Hardened from 1050°C and tempered at 650°C	600	Turbine blades and discs, bolts, some gas-turbine components
0·35 C 1·5 Si 21·0 Cr 7·0 Ni 4·0 W	—	950	Resists a high concentration of sulphurous gases
0·15 C 1·5 Si 25·0 Cr 19·0 Ni	—	1100	Heat-treatment pots and muffles, aircraft-engine manifolds, boiler and super-heater parts
0·35 C 0·6 Si 28·0 Cr	—	1150	Furnace parts, automatic stokers, retorts. Resistant to sulphurous gases

Table 13·9—Heat-resisting steels.

(1) it must resist oxidation and also attack by other gases in the working atmosphere,

(2) it must be strong enough at the working temperature.

13.61 Resistance to oxidation is effected by adding chromium and sometimes small amounts of silicon. Both of these elements coat the surface with a tenacious layer of oxide, which protects the metal beneath from further attack. Nickel toughens the alloy by restricting grain-growth, but increased strength at high temperatures is achieved by adding small amounts of tungsten, titanium, or niobium. These form small particles of carbide, which raise the limiting creep stress (4.51) at the working temperature. Such steels are used for exhaust valves of internal-combustion engines, conveyor chains and other furnace parts, racks for enamelling stoves, annealing-boxes, rotors for steam and gas turbines, and retorts.

Summary of the principal effects of the main alloying elements
13.70 The more important effects of the main alloying elements added to steels can be summarised as follows.

Element	Chemical symbol	Principal effects when added to steel (more important effects in italics)
Manganese	Mn	*Acts as a deoxidiser and a desulphuriser.* Stabilises carbides.
Nickel	Ni	*Toughens steel by refining grain.* Strengthens ferrite. Causes cementite to decompose—hence used by itself only in low-carbon steel.
Chromium	Cr	Stabilises carbides, and forms hard chromium carbide—hence *increases hardness of steel.* Promotes grain-growth, and so causes brittleness. *Increases resistance to corrosion.*
Molybdenum	Mo	*Reduces 'temper brittleness'* in nickel–chromium steels. Stabilises carbides. Improves high-temperature strength.
Vanadium	V	Stabilises carbides. *Raises softening temperature of hardened steels* (as in high-speed steels).
Tungsten	W	*Forms very hard stable carbides. Raises the softening temperature, and renders transformations very sluggish* (in high-speed steels). Reduces grain growth. Raises the limiting creep stress at high temperatures.

13.71 Remember that *all* of these elements increase the depth of hardening of a steel; so an alloy steel has a bigger 'ruling section' than a plain-carbon steel. This is because alloying elements *slow down* the austenite → martensite transformation rates, so it is possible to oil-harden or, in some cases, air-harden a suitable steel.

Chapter Fourteen
The Surface-hardening of Steels

14.10 Many metal components require a combination of mechanical properties which at first sight seems impossible to attain. Thus, bearing metals (19.60) must be both hard and, at the same time, ductile; whilst many steel components, like cams and gears, need to be strong and shock-resistant, yet also hard and wear-resistant. In ordinary carbon steels, these two different sets of properties are found only in materials of different carbon-content. Thus, a steel with about 0·1 % carbon will be tough, whilst one with 0·9 % carbon will be very hard when suitably heat-treated.

The problem can be overcome by using one of two different methods:

(1) by employing a tough, low-carbon steel, and altering the composition of its surface, either by case-hardening or by nitriding;

(2) by using a steel of uniform composition throughout, but containing at least 0·4 % carbon, and heat-treating the surface differently from the core, as in flame- and induction-hardening.

In the first case, it is the hardening material which is localised; whilst in the second case it is the heat-treatment which is localised.

Case-hardening
14.20 This process makes use of the fact that carbon will dissolve in appreciable amounts in *solid* iron, provided that the latter is in the face-centred cubic crystal form. This is due to the fact that carbon dissolves

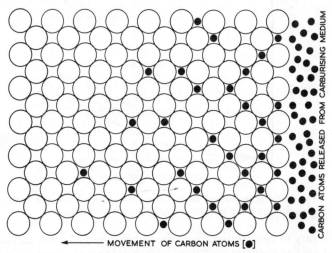

Fig. 14.1—An impression of the penetration by carbon atoms into the lattice structure of FCC iron (austenite).

interstitially in iron (8.34)—the carbon atoms are small enough to infiltrate between the larger iron atoms; so solid iron can absorb carbon in much the same way that water is soaked up by a sponge. Since only face-centred cubic iron will dissolve carbon, it follows that steel must be carburised at a temperature *above* the upper critical temperature. As it is generally mild steel which is carburised, this involves using a temperature in the region of 900–950°C. Thus carburising consists of surrounding mild steel components with some carbon-rich material, and heating them above their upper critical temperature for long enough to produce a carbon-rich surface layer of sufficient depth.

Solid, liquid, and gaseous carburising materials are used, and the quantity of output required largely governs the method employed.

14.21 *Carburising in solid media.* So-called 'pack carburising' is probably the process with which the reader is most likely to be familiar. Components to be treated are packed into steel boxes, along with the carburising-material, so that a space of roughly 50 mm exists between them. Lids are then fixed on the boxes, which are then slowly heated to the carburising temperature (900–950°C). They are maintained at this temperature for up to six hours* according to the depth of case required (fig. 14.2). Carburising-

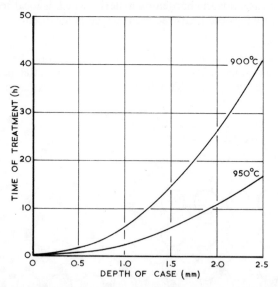

Fig. 14.2—The relationship between time of treatment, temperature, and depth of case in a carburising process using solid media (0·15% plain-carbon steel).

mixtures vary in composition, but consist essentially of some carbon-rich material, such as charcoal or charred leather, along with an energiser which

* Much longer periods are sometimes necessary when deep cases are to be produced.

may account for about 40% of the total. This energiser is generally a mixture of sodium carbonate ('soda ash') and barium carbonate. Its function is to accelerate the solution of carbon by taking part in a chemical reaction which causes single carbon atoms to be released at the surface of the steel.

If it is necessary to prevent any parts of the surface of the component from becoming carburised, this can be achieved by electroplating these areas with copper, to a thickness of 0·07 to 0·10 mm, since carbon does not dissolve in solid copper. In small-scale treatment, the same objective can be achieved by coating the necessary areas of the components with a paste of fireclay and ignited asbestos mixed with water. This is allowed to dry on the surface, before the components are loaded into the carburising-box.

When carburising is complete, the charge is either quenched or allowed to cool slowly in the box, depending on the subsequent heat-treatment it will receive.

14.22 *Carburising in liquid media.* Liquid-carburising—or cyanide-hardening, as it is usually called—is carried out in baths of molten salt which contain 20 to 50% sodium cyanide, together with as much as 40% sodium carbonate, and varying quantities of sodium or barium chloride. The cyanide-rich mixture is heated in iron pots to a temperature of 870 to 950°C, and the work, which is carried in wire baskets, is immersed for periods of about five minutes upwards, according to the depth of case required. The process is particularly suitable for producing shallow cases of 0·1 to 0·25 mm.

Carburising takes place due to the decomposition of sodium cyanide at the surface of the steel. Atoms of both carbon and nitrogen are released; so cyanide-hardening is due to the absorption of nitrogen, as well as of carbon.

The main advantages of cyanide hardening are:

(1) the temperature of a liquid salt bath is uniform throughout, and can be controlled accurately by pyrometers,
(2) the basket of work can be quenched direct from the bath,
(3) the surface of the work remains clean.

Many readers will be aware of the fact that all cyanides are extremely poisonous chemicals. However, since *sodium cyanide is one of the most deadly poisonous materials* in common use industrially, it might be well to stress the following points, which should be observed by the reader should he find himself involved in the use of cyanides.

(1) Every pot should be fitted with an efficient fume-extraction system.
(2) The consumption of food by operators whilst working in a shop containing cyanide should be *absolutely forbidden.*
(3) Cyanide-rich salts should never be allowed to come into contact with an open wound.

(4) Advice should be sought before disposing of any waste hardening salts. They should *never* be tipped into canals or rivers.

14.23 *Carburising by gaseous media.* Gas-carburising is carried out in both continuous and batch-type furnaces. Whichever is used, the components are heated at about 900°C for three hours or more in an atmosphere containing gases which will deposit carbon atoms at the surface of the components. The gases generally used are the hydrocarbons methane (23.41) and propane. These should be of high purity, otherwise oily soot may be deposited on the work-pieces. The hydrocarbon is usually mixed with a 'carrier' gas (generally a mixture of nitrogen, hydrogen, and carbon monoxide), which allows better gas circulation, and hence greater uniformity of treatment.

Gas-carburising is becoming increasingly popular, particularly for the mass production of thin cases. Its main advantages as compared with other methods of carburising are:

(1) the surface of the work is clean after treatment,
(2) the necessary plant is more compact for a given output,
(3) the carbon-content of the surface layers can be more accurately controlled by this method.

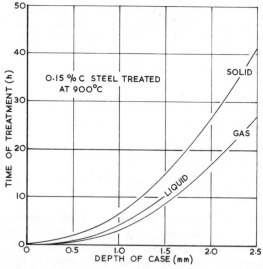

Fig. 14.3—The relationship between time of treatment and depth of case produced when carburising in solid, liquid, and gaseous media (0·15% plain-carbon steel).

Heat-treatment after carburising
14.30 If the carburising process has been successful, the core will still have a low carbon-content (0·1 to 0·3% carbon); whilst the case should have a maximum carbon-content of 0·83% carbon (the eutectoid composition).

Unfortunately, prolonged heating in the austenitic range will have caused the formation of coarse grain, and further heat-treatment is desirable if optimum properties are to be obtained.

The most common method of producing a fine-grained structure in steel is by normalising it. This involves heating the steel to just above its upper critical temperature, followed by cooling it in air (11.51). The need for such treatment poses a problem here, since core and case are of widely different carbon-contents, and therefore have different upper critical temperatures. Thus, if the best mechanical properties are to be obtained in both core and case, a double heat-treatment is necessary.

14.31 *Refining the core.* The component is first heat-treated to refine the grain of the core, and so toughen it. This is done by heating the component to a temperature just above the upper critical temperature for the core (point *A* in fig. 14.4), so that the coarse ferrite/pearlite will be replaced by fine-grained austenite. The component is then generally water-quenched, so that a mixture of fine-grained ferrite and a little martensite is produced. The temperature of this treatment is high above the upper critical temperature for the case (723°C); so at this stage the case will be of coarse-

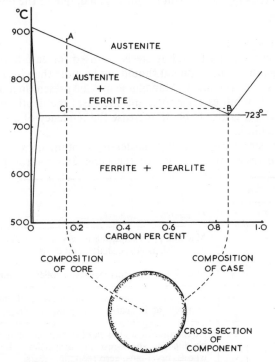

Fig. 14.4—Heat-treatment of a carburised component in relation to the equilibrium diagram.

grained martensite (since the steel was quenched). Further heat-treatment is therefore necessary to refine the grain of the case.

14.32 *Refining the case.* The component is now heated to 760°C (point *B* in fig. 14.4), so that the structure of the case changes to fine-grained austenite. Quenching then gives a hard case of fine-grained martensite.

At the same time, any brittle martensite present in the core as a result of the *first* quenching process will be tempered to some extent by the second heating-operation (point *C* in fig. 14.4). Finally, the component is tempered at 200°C, to relieve any quenching-stresses present in the case.

The above heat-treatment processes can be regarded to some extent as the counsel of perfection, and the needs of economy often demand that such treatments may be replaced by a single operation. Often the work may be 'pot-quenched'; that is, quenched direct from the carburising process, followed by a low-temperature tempering process to relieve any quenching-stresses.

Alternatively, the work may be cooled slowly from the carburising temperature, to give maximum ductility to the core. It is then reheated to 760°C, and water-quenched. This treatment leaves the core quite soft, but hardens the case, which will be fine-grained, due to the low quenching temperature.

Case-hardening steels
14.40 Plain-carbon and low-alloy steels are used for case-hardening, but, in either type, the carbon-content should not be more than 0·25% if a really tough core is to be obtained. Manganese may be present in amounts up to 0·9%, since it stabilises cementite, and increases the depth of hardening. Unfortunately, it is also liable to increase the tendency of a steel to crack during quenching.

Low-alloy steels used for case-hardening contain up to 5·0% nickel, since this increases the strength of the core, and retards grain-growth

Typical composition %				Characteristics and uses
C	Mn	Ni	Cr	
0·15	0·7	—	—	Machine parts requiring a very hard surface and a tough core, such as gears, shafts, and cams.
0·25	0·5	—	—	Mainly for high-duty ball- and roller-bearings.
0·15	0·5	3·0	—	Used where combined hardness and toughness are required, as in gears, crank-pins, coupling-pins, etc.
0·15	0·4	4·0	1·2	Parts requiring a glass-hard surface combined with stress-bearing and shock-resisting properties. Used for crown-wheels, bevel-pins, aero reduction-gears.

Table 14.1—Case-hardening steels.

during the carburising process. This often means that the core-refining heat-treatment can be omitted. Chromium is sometimes added to increase hardness and wear-resistance of the case, but it must be present only in small quantities, as it tends to promote grain-growth (13.22).

Nitriding

14.50 Nitriding and case-hardening have one factor in common—both processes involve heating the steel for a considerable time in the hardening-medium; but, whilst in case-hardening the medium contains carbon, in nitriding it contains gaseous nitrogen. Special steels—'Nitralloy' steels—are necessary for the nitriding process, since hardening depends upon the formation of very hard compounds of nitrogen and such metals as aluminium, chromium, and vanadium. Ordinary plain-carbon steels cannot be nitrided, since any compounds of iron and nitrogen which form will diffuse into the core, so that the increase in hardness of the surface is lost. The hard compounds formed by aluminium, chromium, and vanadium, however, remain near to the surface, providing an extremely hard skin.

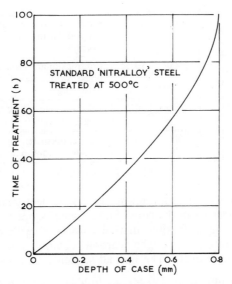

Fig. 14.5—The relationship between time of treatment and depth of case produced in the nitriding process.

14.51 Nitriding is carried out at the relatively low temperature of 500°C. Consequently, it is made the *final* operation in the manufacture of the component, all machining and core heat-treatments having been carried out previously. The work is maintained at 500°C for between forty and one hundred hours, according to the depth of case required, though treatment for ninety hours is general. The treatment takes place in a gas-tight chamber through which ammonia gas is allowed to circulate. Some of the ammonia decomposes, releasing single atoms of nitrogen, which are at once absorbed by the surface of the steel. Ordinary 'atmospheric' nitrogen is not suitable, since it exists in the form of molecules, which would not be absorbed by the steel.

14.52 Nitralloy steels containing aluminium are hardest, since aluminium forms very hard compounds with nitrogen. Unfortunately, aluminium tends to affect the core-strength adversely, and is replaced by chromium, vanadium, and molybdenum in those Nitralloy steels in which high strength and toughness of the core are important. Compositions and uses of some nitriding steels are given in table 14.2.

Composition %					Typical mechanical properties		Characteristics and uses
C	Cr	Mo	V	Al	Tensile strength (N/mm²)	VPN	
0·5	1·5	0·2	—	1·1	1200	1075	Where maximum surface hardness, coupled with high core-strength is essential
0·2	1·5	0·2	—	1·1	600	1075	For maximum surface hardness, combined with ease of machining before hardening
0·4	3·0	1·0	0·2	—	1400	875	Ball-races, etc. where high core-strength is necessary
0·3	3·0	0·4	—	—	1000	875	Aero crankshafts, air-screw shafts, aero cylinders, crank-pins, and journals

Table 14.2—Nitriding steels.

14.53 Prior to being nitrided, the work-pieces are heat-treated, to produce the required properties in the core. Since greater scope is possible in this heat-treatment than is feasible in that associated with case-hardening, Nitralloy steels often have higher carbon-contents, allowing high core-strengths to be developed. The normal sequence of operations will be:

(1) oil-quenching from 850–900°C, followed by tempering at between 600 and 700°C;

(2) rough machining, followed by a stabilising anneal at 550°C for five hours, to remove internal stresses;

(3) finish-machining, followed by nitriding.

Any areas of the surface which are required soft are protected by coating with solder or pure tin, by nickel-plating, or by painting with a mixture of whiting and sodium silicate.

14.54 *Advantages of nitriding* over case-hardening are as follows.

(1) Since no quenching is required *after* nitriding, cracking or distortion is unlikely, and components can be machine-finished before treatment.

(2) An extremely high surface hardness of up to 1150 VPN is attainable with the aluminium-type Nitralloy steels.

(3) Resistance to corrosion is good, if the nitrided surface is left unpolished.

(4) Hardness is retained up to 500°C; whereas a case-hardened component begins to soften at about 200°C.

(5) The process is clean, and simple to operate.

(6) It is cheap if large numbers of components are to be treated.

14.55 *Disadvantages of nitriding* as compared with case-hardening are as follows.

(1) The initial outlay for nitriding plant is higher than that associated with solid- or liquid-medium carburising; so nitriding is only economical when large numbers of components are to be treated.

(2) If a nitrided component is accidentally over-heated, the loss of surface hardness is permanent, unless the component can be nitrided again. A case-hardened component would need only to be heat-treated, assuming that it had not been so grossly over-heated as to decarburise it.

Carbonitriding

14.60 This is a surface-hardening process which makes use of a mixture of hydrocarbons and ammonia. It is therefore a gas treatment, and is sometimes known as 'dry-cyaniding'—a reference to the fact that a mixed carbide-nitride case is produced, as in ordinary liquid-bath cyanide processes (14.22).

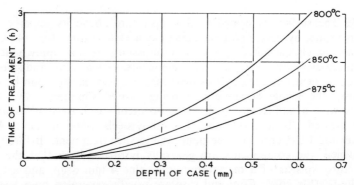

Fig. 14.6—The relationship between time of treatment, temperature, and depth of case produced by the carbonitriding process.

14.61 Furnaces used for carbonitriding are generally of the continuous type, as the work is nearly always directly quenched in oil from the carbonitriding atmosphere. If 'stopping off' is necessary for any areas required soft, then good quality copper-plating is recommended.

Carbonitriding is an ideal process for hardening small components where great resistance to wear is necessary.

Flame-hardening

14.70 In this process, the work-piece is of uniform composition throughout, and it is the type of structure which varies across the section, because the surface layers have received extra heat-treatment as compared with the core material.

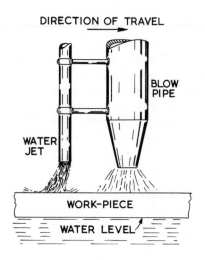

DIRECTION OF TRAVEL

BLOW PIPE

WATER JET

WORK-PIECE

WATER LEVEL

Fig. 14.7—The principles of flame-hardening.

14.71 The surface is heated to a temperature above its upper critical temperature, by means of a travelling oxy-acetylene torch (fig. 14.7), and is immediately quenched by a jet of water issuing from a supply built into the torch-assembly. Symmetrical components, such as gears and spindles, are conveniently treated by this process, since they can be spun between centres, the whole circumference being treated simultaneously. Only steels with a sufficiently high carbon-content—at least 0·4%—can be hardened effectively in this way. Low-alloy steels containing up to 4·0% nickel and 1·0% chromium respond well to such treatment. Before being hardened, the components are generally normalised, so that the final structure consists of a martensitic case some 4 mm deep, and a tough ferrite-pearlite core. Core and case are usually separated by a layer of bainite, which helps to prevent the hard case from cracking away from the core material. Should a final tempering process be necessary, this can also be carried out by flame-heating, though furnace treatment is also possible, since such low-temperature treatment will have no effect on the core, particularly if it has been normalised.

Induction-hardening

14.80 This process is similar in principle to flame-hardening, except that the component is usually held stationary whilst the whole circumference is heated simultaneously by means of an induction-coil. This coil carries a high-frequency current, which produces eddy currents in the surface of the component, thus raising its temperature. The depth to which heating occurs varies inversely as the square root of the frequency; so that the higher the frequency used, the shallower the depth of heating. Typical frequencies used are:

> 3000 Hz for depths of 3 to 6 mm,
> 9600 Hz for depths of 2 to 3 mm.

As soon as the surface of the component has reached the necessary quenching temperature, the current is switched off, and the surface is simultaneously quenched by pressure jets of water, which pass through holes in the induction-block (fig. 14.8).

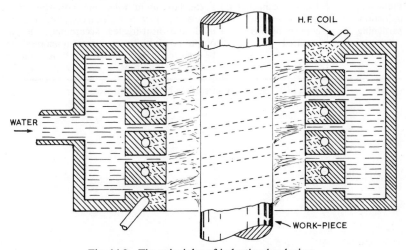

Fig. 14.8—The principles of induction-hardening.

This process lends itself to mechanisation; so that selected regions of a symmetrical component can be hardened, whilst others are left soft. As in flame-hardening, the induction process makes use of the existing carbon-content—which consequently must be at least 0·4%—whilst in case-hardening, nitriding, and carbonitriding, an alteration in the composition of the surface layers takes place.

Summary of surface-hardening processes
14.90 The characteristics and uses of the processes dealt with in this chapter are summarised in table 14.3.

Process	Type of work	Characteristics
Case-hardening (solid and gas)	Gears, king-pins, ball- and roller-bearings, rocker-arms, gauges	A wide variety of low-carbon and low-alloy steels can be treated. Local soft surfaces are easily retained. Gas carburising is a rapid process.
Case-hardening (liquid cyanide)	Used mainly for light cases	The case tends to be of poorer quality, but thin cases can be produced quickly.
Nitriding	Crankshafts, cam-shafts, gears requiring high core-strength	A very high surface hardness, combined with a high core-strength when required. Surface will withstand tempering influences up to 500°C. Less suitable than other methods if surface has to withstand very high pressure, e.g. gear-teeth.
Carbo-nitriding	Particularly useful for treating small components	Safe, clean, and easy to operate, applicable to mass-production methods.
Flame- and induction-hardening	Tappets, cam-shafts, gears where high core-strength is required	Particularly useful where high core-strength is necessary, since a high-carbon steel can be used and heat-treated accordingly. Rapid output possible, but equipment often needs to be designed for a particular job; hence suitable mainly for long runs.

Table 14.3—Summary of surface-hardening processes.

Chapter Fifteen
Cast Iron

15.10 The Victorian era may well be remembered by the cast-iron monstrosities which it produced. Street lamps, domestic fireplaces, and railings were typical cast-iron products of that period. Most of these relics are gone—the railings fell victim of the need for steel during the Second World War—but many an industrial town still boasts an ornamental drinking-fountain in its local park, or a cast-iron clock presiding over the public-conveniences (of similar period) in the town square.

During the nineteenth century, much cast iron was also used for engineering purposes. Today, the whole production of cast iron is directed towards these purposes, and, as in other fields of metallurgical technology, considerable progress has been made during the present century. Special high-duty and alloy compositions have made cast iron an extremely important engineering material, which is suitable for the manufacture of crankshafts, connecting rods, and axles—components which were formerly made from forged steel.

15.11 Ordinary cast iron is similar in composition to the crude pig iron produced by the blast-furnace. The pig iron is generally melted in a cupola, any necessary adjustments in composition being made during the melting process. At present, the high cost of metallurgical coke, coupled with the desire to produce high-grade material, has led the foundryman to look for other methods of melting cast iron; consequently, line-frequency induction furnaces are being used on an increasing scale.

15.12 The following features make cast iron an important material.

(1) It is a cheap metallurgical substance, since it is produced by simple adjustments to the compositions of ordinary pig irons.

(2) Mechanical rigidity and strength under compression are good.

(3) It machines with ease when a suitable composition is selected.

(4) Good fluidity in the molten state leads to the production of good casting-impressions.

(5) High-duty cast irons can be produced by further treatment of irons of suitable composition; e.g. spheroidal-graphite irons are strong, whilst malleable irons are tough.

The composition of cast iron
15.20 Ordinary cast irons contain the following elements:

carbon	3·0–4·0%	sulphur	up to 0·1%
silicon	1·0–3·0%	phosphorus	up to 1·0%
manganese	0·5–1·0%		

15.21 *Carbon* may be present in the structure either as flakes of graphite or as a network of hard, brittle iron carbide (or cementite). Naturally, if a cast iron contains much of this brittle cementite, its mechanical properties will be poor, and for most engineering purposes it is desirable for the carbon to be present as small flakes of graphite. Cementite is a silvery-white compound, and, if an iron containing much cementite is broken, the fractured surface will be silvery white, because the piece of iron breaks along the brittle cementite networks. Such an iron is termed a *white iron*. Conversely, if an iron contains much graphite, its fractured surface will be grey, due to the presence of graphite, and this iron would be called a *grey iron*.

15.22 *Silicon* to some extent governs the form in which carbon is present in cast iron. It causes the cementite to be unstable, so that it decomposes, thus releasing free graphite. Therefore a high-silicon iron tends to be a grey iron, whilst a low-silicon iron tends to be a white iron (fig. 15.1).

15.23 *Sulphur* has the opposite effect on the structure; that is, it tends to stabilise cementite, and so helps to produce a white iron. However, sulphur causes excessive brittleness in cast iron (as it does in steel), and it is therefore always kept to the minimum amount which is economically possible. No self-respecting foundryman would think of altering the structure of an iron by adding sulphur to a cupola charge, any more than he would think of diluting his whisky with dirty washing-up water; instead, he obtains the desired microstructure by adjusting the silicon-content of the iron.

During the melting of cast iron in a cupola, some silicon is inevitably burned away, whilst some sulphur will be absorbed from the coke. Both factors tend to make the iron 'whiter', so the foundryman begins with a charge richer in silicon than that with which he expects to finish.

15.24 *Manganese* toughens and strengthens an iron, partly because it neutralises much of the unwelcome sulphur by forming a slag with it, and partly because some of the manganese dissolves in the ferrite.

15.25 *Phosphorus* forms a very brittle compound with some of the iron; it is therefore kept to a minimum amount in most engineering cast irons. However, like silicon, it increases fluidity, and considerably improves the casting qualities of irons which are to be cast in thin sections, assuming that components are involved in which mechanical properties are unimportant. Thus cast-iron water pipes contain up to 0·8% phosphorus, whilst many of the old ornamental castings contained up to 1·0% of the element.

The influence of cooling rate on the properties of a cast iron
15.30 When the presence of silicon in an iron tends to make cementite unstable, the latter does not break up or decompose instantaneously: this

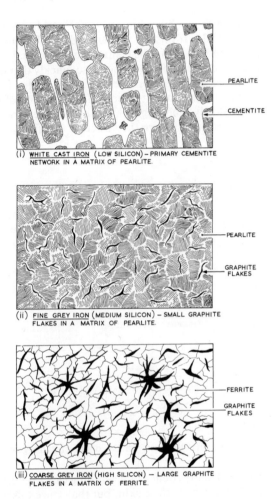

(i) <u>WHITE CAST IRON</u> (LOW SILICON) – PRIMARY CEMENTITE
NETWORK IN A MATRIX OF PEARLITE.

(ii) <u>FINE GREY IRON</u> (MEDIUM SILICON) – SMALL GRAPHITE
FLAKES IN A MATRIX OF PEARLITE.

(iii) <u>COARSE GREY IRON</u> (HIGH SILICON) – LARGE GRAPHITE
FLAKES IN A MATRIX OF FERRITE.

Fig. 15.1—The effects of silicon-content on the structure of cast iron.
The higher the silicon-content, the more unstable the cementite becomes, until even
the pearlitic cementite decomposes (iii). Magnifications—approx. $100 \times$.

process of decomposition requires time. Consequently, if such an iron is cooled so that it solidifies rapidly, the carbon may well be 'trapped' in the form of hard cementite, and so give rise to a white iron. On the other hand, if this iron is allowed to cool and solidify slowly, the cementite has more opportunity to decompose, forming graphite, and so produce a grey iron.

15.31 This effect can be shown by casting a 'wedge-bar' in an iron of suitable composition (fig. 15.2). If this bar is fractured, and hardness determinations are made at intervals along the centre line of the section, it will be found that the thin end of the wedge has cooled so quickly that decomposition of the cementite has not been possible. This is indicated by the white fracture and the high hardness in that region. The thick end of the wedge, however, has cooled slowly, and is graphitic, because cementite has had more opportunity to break up. Here the structure is softer.

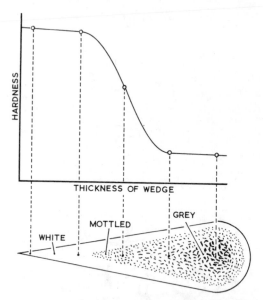

Fig. 15.2—The effect of sectional thickness on the depth of chilling of a grey iron.

15.32 If the reader has had experience in machining cast iron, he will know that such a casting has a hard surface skin, but that, once this is removed, the material beneath is easy to machine. This hard skin consists largely of cementite which has been prevented from decomposing by the chilling action of the mould. Iron beneath the surface has cooled more slowly, so that cementite has decomposed, releasing graphite.

15.33 To summarise: the engineer requires a cast iron in which carbon is present in the form of small flakes of graphite. The form in which the carbon is present depends upon:

(1) the silicon-content of the iron, and

(2) the rate at which the iron solidifies and cools, which in turn, depends upon the cross-sectional thickness of the casting.

Thus the foundryman must strike a balance between the silicon-content of the iron and the rate at which it cools.

SILICON INCREASES ↑ — GREY IRON ⎡COARSE GRAPHITE (RELATIVELY WEAK)
⎣FINE GRAPHITE (TOUGH AND STRONG)

WHITE IRON – CEMENTITE (WEAK AND BRITTLE) ↓ COOLING RATE INCREASES

When casting *thin* sections, it will be necessary for the foundryman to choose an iron which has a coarser grey fracture in the pig form than that which is required in the finished casting. Thus the iron must have a higher silicon-content than that used for the manufacture of castings of heavier section, which will consequently cool more slowly.

15.34 Sometimes it is necessary to have a hard-wearing surface of white iron at some point in a casting which otherwise requires a tough grey-iron structure. This can be achieved by incorporating 'chills' at appropriate points in the sand mould. The 'chill' usually consists of a metal block, which will cause the molten iron in that region to cool so quickly that a layer of hard cementite is retained adjacent to the chill.

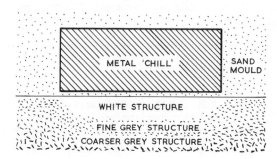

METAL 'CHILL' SAND MOULD

WHITE STRUCTURE

FINE GREY STRUCTURE

COARSER GREY STRUCTURE

Fig. 15.3—The use of 'chills' in iron-founding.

'Growth' in cast irons

15.40 An engineering cast iron contains some cementite as a constituent of the pearlitic areas of the structure. If such an iron is heated for a prolonged period in the region of 700°C or above, this cementite decomposes to form graphite and iron. Since the graphite and iron so formed occupy more space than did the original cementite, the volume of the heated region increases, and this expansion leads to warping of the casting, and the formation of cracks at the surface. Hot gases penetrate these cracks, gradually oxidising both the graphite and the iron, so that the surface ultimately disintegrates. Fire bars in an ordinary domestic grate often break up in this way.

15.41 The best way to prevent 'growth' in cast irons which are to be used at high temperatures is to ensure that they contain *no* cementite in the first place. This can be achieved by using a high silicon-content iron, in which decomposition of all the cementite will take place during the actual solidification of the casting. Thus, 'Silal' contains 5·0% silicon, with relatively low carbon, so that the latter is present in the finished casting entirely as graphite. Unfortunately, Silal is rather brittle; so, when the cost is justified, the alloy cast iron 'Nicrosilal' (table 15.3) may be used.

Ordinary cast irons
15.50 Ordinary cast irons fall into two main groups.

15.51 *Engineering irons*, which must possess reasonable strength and toughness, generally coupled with good machinability (21.21). The silicon-content of such an iron will be chosen in accordance with the cross-sectional thickness of the casting to be produced. It may be as much as 2·5% for castings of thin section, but as low as 1·2% for bulky castings of heavy section. This relationship between silicon-content and sectional thickness of a casting is illustrated by the three irons specified for light, medium, and heavy machine castings given in table 15.1. Amounts of sulphur and phosphorus generally are kept low, since both elements cause brittleness, though some castings contain as much as 1·0% phosphorus, to give fluidity to the molten iron.

Composition %					Uses
C	Si	Mn	S	P	
3·30	1·90	0·65	0·08	0·15	Motor brake-drums
3·25	2·25	0·65	0·10	0·15	Motor cylinders and pistons
3·25	2·25	0·50	0·10	0·35	Light machine-castings
3·25	1·75	0·50	0·10	0·35	Medium machine-castings
3·25	1·25	0·50	0·10	0·35	Heavy machine-castings
3·60	1·75	0·50	0·10	0·80	Water-pipes
3·50	2·75	0·50	0·10	0·90	Low-strength ornamental castings of yesteryear

Table 15.1—Compositions and uses of some ordinary cast irons.

The principal engineering grey irons are covered by BS 1452: 1961, which deals with seven different grades, each with its minimum acceptable tensile strength. Grade 10 refers to a fluid iron which may contain up to 1·0% phosphorus, and which will be used only where negligible strength is required, as, for example, in domestic rain-water pipes and gutters. Grade 12 is a high-silicon iron which may contain up to 3·5% silicon, so that nearly all of the carbon is present as coarse graphite flakes in a background mainly of ferrite. It is relatively weak, because of the presence of large graphite flakes. At the other extreme, grade 26 is a low-silicon iron, containing fine graphite flakes in a matrix of pearlite, and is consequently much stronger. Between these two extremes, differing silicon-contents control the structures and consequently the mechanical properties. Most engineering castings are made from an iron selected from grades 14 to 26.

Grade	10	12	14	17	20	23	26
Tensile strength (min) (N/mm²)	154	185	215	262	308	354	400
Strength in compression (N/mm²)	600	600	775	775	1000	up to	1225
Hardness (Brinell)		155–240			180–250		240–320
Impact toughness (J) of an un-notched test piece approx. 20 mm in diameter	11	11	12	17	24	36	36
			approximate values				

Table 15.2—Typical properties of engineering grey irons in accordance with BS 1452: 1961. (SI equivalents are given for original imperial units.)

15.52 *Fluid irons*, in which mechanical strength is of secondary importance, were at one time widely used in the manufacture of railings, lamp-posts, and fireplaces. High fluidity was necessary in order that the iron should fill intricate mould impressions, and this was achieved by using a high silicon-content of 2·5 to 3·5%, as well as a high phosphorus-content of up to 1·5%. Cast iron has now been replaced by other materials for the purposes mentioned, though the products are often no less hideous in appearance, particularly in so far as reinforced-concrete structures are concerned.

High-duty cast irons
15.60 When we look at the microstructure of an ordinary engineering iron, we see the graphite as long, thin flakes, each terminating in a sharply-pointed end (fig. 15.1). We must remember, however, that we are looking at two-dimensional cross-sections of these flakes, and that graphite is in fact present in the form of thin, three-dimensional plates, with sharp-edged rims. Graphite has no appreciable strength, and so the flakes have the same effect on mechanical properties as would have cavities in the structure. The sharp edges act as stress-raisers within the structure; that

is, they give rise to an increase in local stress, in the same way that a sharp-cornered key-way tends to weaken a shaft. Hence, both the mechanical strength and the toughness of a cast iron can be improved by some treatment which disperses the graphite flakes, or, better still, which alters their shape to that of spherical globules.

15.61 *'Inoculated'* *cast irons* are treated with small amounts of a calcium-silicon compound just before casting. This produces a considerable refinement of the grain, so that the coarse graphite flakes are replaced by much smaller ones, and the mechanical properties are improved as a result. The 'Meehanite' process is a proprietary treatment of this type.

15.62 *Spheroidal-graphite* *('SG')* *cast iron*, also known as 'nodular' iron, or (in the USA) as 'ductile' iron, contains its graphite in the form of rounded globules (fig. 15.4). The sharp-edged, stress-raising flakes are thus eliminated, and the structure is made more continuous. Graphite is made to deposit in globular form by adding small amounts of either of the metals cerium or magnesium to the molten iron, just before casting. Magnesium is the more widely used, and is generally added as a magnesium-nickel alloy, in amounts to give a residual magnesium-content of 0·1% in the iron. Such an iron may have a tensile strength of as much as 775 N/mm².

Malleable cast irons

15.70 These are *white* irons, of suitable composition, which have been cast to shape in the ordinary way, and have then received some form of heat-treatment. The object of this heat-treatment is either to convert the cementite into small spherical particles of carbon (the 'black-heart' process), or, alternatively, to remove the carbon completely from the structure (the 'white-heart' process). In either process, the silicon-content of the iron is usually less than 1·0%, in order that the iron shall be 'white' in the cast condition. When the cementite has been either replaced by carbon or removed completely, a product which is both malleable and ductile is the result.

15.71 *Blackheart malleable iron.* In this process, the low-silicon white-iron castings are heated at about 900°C in a continuous-type furnace through which an oxygen-free atmosphere circulates. A moving hearth carries the castings slowly through the heating-zone, so that the total heating time is about forty-eight hours.

This prolonged annealing causes the cementite to break down; but, instead of coarse graphite flakes, the carbon deposits as small 'rosettes' of 'temper carbon'. A fractured surface appears dark, because of the presence of this carbon; hence the term 'blackheart'. Since the structure now consists entirely of temper carbon and ferrite, it is soft and ductile. Black-

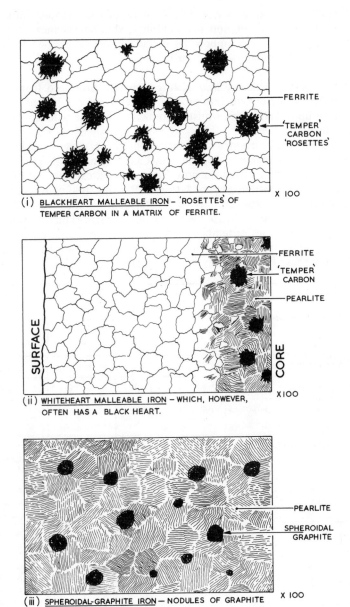

(i) <u>BLACKHEART MALLEABLE IRON</u> – 'ROSETTES' OF TEMPER CARBON IN A MATRIX OF FERRITE.

(ii) <u>WHITEHEART MALLEABLE IRON</u> – WHICH, HOWEVER, OFTEN HAS A BLACK HEART.

(iii) <u>SPHEROIDAL-GRAPHITE IRON</u> – NODULES OF GRAPHITE IN A PEARLITE MATRIX.

Fig. 15.4—Structures of malleable and spheroidal-graphite cast irons.
The structure of a whiteheart malleable iron (ii) is rarely uniform throughout, and often contains some carbon in the core. This has had insufficient time to diffuse outwards to the skin, and so be lost.

heart malleable castings are widely used in the automobile industries, because of their combination of castability, shock-resistance, and good machinability. Typical components include brake-shoes, pedals, levers, wheel-hubs, axle-housings, and door-hinges.

15.72 *Whiteheart malleable iron* castings are also produced from a low-silicon white iron; but, in this process, the castings are heated at about 1000°C for up to 100 hours, whilst in contact with some oxidising material such as haematite ore.

During heating, carbon at the surface of the castings is oxidised by contact with the haematite ore, and is lost as carbon dioxide gas. Carbon then diffuses outwards from the core—rather like a carburising process in reverse—and is in turn lost by oxidation. After this treatment is complete, *thin* sections will consist entirely of ferrite, and, on fracture, will give a steely-white appearance; hence the name 'whiteheart'. Thick sections may contain some particles of temper carbon at the core, because heating has not been sufficiently prolonged to allow all of the carbon to diffuse outwards from these thick sections (fig. 15.4 (ii)).

Name or type of iron	Composition %			
	C	Si	Mn	Others
Chromidium	3·2	2·1	0·8	Cr—0·3
Ni-tensyl	2·8	1·4	—	Ni—1·5
Wear-resistant iron	3·6	2·8	0·6	V—0·2
Ni-Cr-Mo iron	3·1	2·1	0·8	Ni—0·5 Cr—0·9 Mo—0·9
Heat-resistant iron	3·4	2·0	0·6	Ni—0·35 Cr—0·65 Cu—1·25
Ni-hard	3·3	1·1	0·5	Ni—4·5 Cr—1·5
Ni-resist	2·9	2·1	1·0	Ni—15·0 Cr—2·0 Cu—6·0
Nicrosilal	2·0	5·0	1·0	Ni—18·0 Cr—5·0
Silal	2·5	5·0	—	—

Table 15.3—Properties and uses of some alloy cast irons.

The whiteheart process is particularly suitable for the manufacture of thin sections which require high ductility. Pipe-fittings, parts for agricultural machinery, switchgear equipment, and fittings for bicycle and motor-cycle frames are typical of whiteheart malleable castings.

15.73 *Pearlitic malleable iron* is similar in its initial composition to the blackheart material. The castings are malleabilised, either fully or partially at about 950°C, to cause the breakdown of the bulk of the cementite, and are then reheated to about 900°C. This reheating process causes some of the carbon to dissolve in austenite which forms above the lower critical temperature; and, on cooling, a background of pearlite is produced, the mechanism of the process being something like that of normalising in steels.

The choice between ordinary grey iron, spheroidal-graphite iron, and malleable iron is, as with most materials problems, governed by both economic and technical considerations. Grey iron is, of course, cheapest, and also the easiest in which to produce sound castings. For 'high-duty' purposes, the choice between SG iron and malleable iron may be less easy,

Typical mechanical properties		Uses
Tensile strength (N/mm²)	Brinell	
270	230	Cylinder-blocks, brake-drums, clutch-casings, etc.
350	230	An 'inoculated' cast iron
—	—	Piston-rings for aero, automobile and diesel engines. Possesses wear-resistance and long life
360	300	Hard, strong, and tough—used for automobile crankshafts
270	220	Good resistance to wear and to heat cracks—used for brake-drums and clutch-plates
—	600	A 'martensitic' iron, due to the presence of nickel and chromium—used to resist severe abrasion—chute-plates in coke plant
210	130	Plant handling salt water—an austenitic iron
255	330	Also an austenitic corrosion- and heat-resistant iron
165	—	'Growth'-resistant at high temperatures

and a number of factors may have to be considered in arriving at a decision.

Alloy cast irons

15.80 Generally speaking, the effects which alloying elements have on the properties of cast iron are similar to the effects which the same elements have on steel. Alloying elements can therefore be used to improve the mechanical properties of an iron, by refining the grain size, stabilising hard carbides, and, in some cases, producing cast irons with a martensitic or austenitic structure.

15.81 *Nickel*, like silicon, has a graphitising effect on cementite, and so tends to produce a grey iron. At the same time, nickel has a grain-refining effect, which helps to prevent the formation of coarse grain in those heavy sections which cool slowly. It also toughens thin sections, which might otherwise be liable to crack.

15.82 *Chromium* is a carbide stabiliser, and forms chromium carbide, which is harder than ordinary cementite. It is therefore used in wear-resistant irons. Since chromium forms very stable carbides, irons which contain chromium are less susceptible to 'growth'.

15.83 *Molybdenum* increases the hardness of thick sections, and also improves toughness.

15.84 *Vanadium* increases both strength and hardness; but, more important still, promotes heat-resistance in cast irons, by stabilising carbides so that they do not decompose on heating.

15.85 *Copper* dissolves in iron in only very small amounts, and has little effect on mechanical properties. It is added mainly to improve resistance to rusting.

Some typical alloy cast irons are shown in table 15.3.

Chapter Sixteen
Copper and its Alloys

16.10 History books generally tell us that the Stone Age was followed, some 3500 years ago, by the Bronze age. It seems fairly certain, however, that metallic copper was used in Egypt at least 1500 years before bronze, but, since copper is the more easily corroded of the two materials, it is not surprising that archaeologists have only limited evidence of this earlier use of copper. Nevertheless, we can regard copper—the third metal of the 'Big Three'*—as Man's most ancient metallic material.

Until recent years, copper ranked second only to iron in terms of world output. Increases in production costs of the metal have driven the engineer to seek cheaper alternatives to copper, and this, in turn, has led to an increase in the output of aluminium, which has now stolen second place from copper in terms of world output.

16.11 In former days, the bulk of the world's requirement of copper was smelted in Swansea, from ore mined in Cornwall, Wales, or Spain. Later, deposits of ore were discovered in the Americas and Australia, and shipped to Swansea to be smelted, but it was subsequently realised that it would be far more economical to smelt the ore at the mine. Thus, Britain ceased to be the centre of the copper industry. Today, the USA—along with Chile, Canada, Zambia, and Katanga—is the leading producer of copper.

The extraction of copper
16.20 Copper is extracted almost entirely from ores based on copper pyrites (a mineral in which copper is chemically combined with iron and sulphur). The metallurgy of the process is rather complex, but is essentially as follows.

(1) The ore is 'concentrated'; that is, it is treated by 'wet' processes to remove as much as possible of the earthy waste, or 'gangue'.

(2) The concentrate is then heated in a current of air, to burn away much of the sulphur. At the same time, other impurities, such as iron and silicon, oxidise to form a slag which floats on top of the purified molten copper sulphide (called 'matte').

(3) The molten matte is separated from the slag, and treated in a Pierce-Smith converter, the operation of which resembles to some extent that of the Bessemer converter formerly used in the manufacture of steel. Some of the copper sulphide is oxidised, and the copper oxide thus formed reacts chemically with the remainder of the sulphide, producing crude copper.

* Iron, aluminium, and copper.

16.21 The crude copper is then refined by either

(1) remelting it in a furnace, so that impurities are oxidised, and are lost as a slag; or

(2) electrolysis, in which an ingot of crude copper is used as the anode, whilst a thin sheet of pure copper serves as the cathode (fig. 16.1). During electrolysis, the anode gradually dissolves, and high-purity copper is deposited on the cathode. 'Cathode copper' so formed is 99·97% pure.

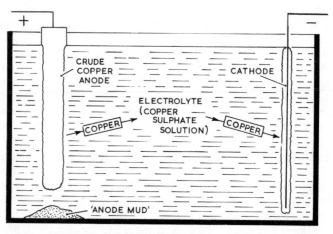

Fig. 16.1—The electrolytic purification of copper.

The 'anode mud' consists of impurities from the copper anode, and often contains sufficient gold and silver to make their recovery worthwhile. During electrolysis, copper atoms are dissolved from the anode, and then deposited on the cathode.

Properties of copper

16.30 The most important physical property of copper is its *very high electrical conductivity*. In this respect it is second only to silver; though, if we take the electrical conductivity of silver as being 100 units, then that of pure copper reaches 97. Consequently, the greater part of the world's production of metallic copper is used in the electrical industries.

Much of the copper used for electrical purposes is of very high purity, as the presence of impurities reduces the electrical conductivity, often very seriously. Thus, the introduction of only 0·04% phosphorus will reduce the electrical conductivity by almost 25%. Other elements have less effect; for example, 1·0% cadmium, added to copper used for telephone-wires, in order to strengthen them, has little effect on the conductivity.

16.31 The thermal conductivity and corrosion-resistance of copper are also high, making it a useful material for the manufacture of radiators, boilers, and other heating equipment. Since copper is also very malleable and ductile, it can be rolled, drawn, deep-drawn, and forged with ease.

In recent years, the cost of copper production has risen steeply; so for many purposes—electrical and otherwise—it has been replaced by aluminium, even though the electrical and thermal conductivities of the latter (17.32) are inferior to those of copper.

Commercial grades of copper

16.40 Copper is refined either electrolytically or by furnace-treatment, as outlined above, and both varieties are available commercially.

16.41 *Oxygen-free high-conductivity* (*OFHC*) *copper* is derived from the electrolytically refined variety. The cathodes are melted, cast, rolled, and then drawn to wire or strip for electrical purposes. This grade is usually 99·97% pure, and is of the highest electrical conductivity.

16.42 '*Tough-pitch*' *copper* is a fire-refined variety which contains small amounts of copper oxide as the main impurity. Since this oxide is present in the microstructure as tiny globules which have little effect on the properties, tough-pitch copper is suitable for many purposes where only moderate electrical and thermal conductivities are required. It is unsuitable for flame-welding processes, because reactions between the oxide globules and the welding flame cause unsoundness, due to the formation of blowholes.

16.43 *Deoxidised copper* is made from the tough-pitch grade by treating it with a small amount of phosphorus just before casting, in order to remove the oxide globules. Whilst phosphorus-deoxidised copper may be valuable for processes where welding is involved, it is definitely not suitable for electrical purposes, because of the big reduction in electrical conductivity introduced by the presence of dissolved phosphorus.

The tensile strength of hard-rolled copper reaches about 375 N/mm²; so, for most engineering purposes where greater strength is required, copper must be suitably alloyed.

Alloys of copper

Alloys of copper are probably less widely used than they were at one time. The relatively high rate of increase in the price of copper, coupled with the fact that the quality of cheaper alternative materials has improved in recent years, has led to the replacement of copper alloys for many purposes. Moreover, improved shaping techniques have allowed less ductile materials to be employed; thus deep-drawing quality mild steel is now often used where ductile brass was once considered to be essential. Nevertheless, the following copper alloys are still of considerable importance.

The brasses

16.50 These are copper-base alloys containing up to 45% zinc and, sometimes, small amounts of other metals, the chief of which are tin, lead, aluminium, manganese, and iron. The equilibrium diagram (fig. 16.2) shows that plain copper-zinc alloys with up to approximately 37% zinc

BS 2870 designation	Composition %			Condition
	Cu	Zn	Other metals	
CZ 101	90	10	—	Soft Hard
CZ 106	70	30	—	Soft Hard
CZ 107	65	35	—	Soft Hard
CZ 108	63	37	—	Soft Hard
CZ 123	60	40	—	Hot-rolled
CZ 121	58	39	Lead—3	Extruded rod
CZ 112	62	37	Tin—1	Extruded
CZ 114	58	Rem	Up to 7% of Al, Fe, Sn, Pb combined	Class A Class B

Table 16.1—Brasses.

have a structure consisting of a single phase—that labelled α. Any substance (or 'phase') which is represented by the Greek letter α, and which is indicated in a position like this on the extreme edge of an equilibrium diagram, is a solid solution.* Solid solutions are generally tough and ductile, and this particular one is no exception, being the basis of one of the most malleable and ductile metallurgical materials in common use. Brasses containing between 10 and 35% zinc are widely used for deep-drawing and general presswork—the maximum ductility being attained in the case of 70–30 brass, commonly known as 'cartridge metal', since it is used in the deep-drawing of cartridge- and shell-cases of all calibres.

* The α (ferrite) phase in the iron-carbon system is an example already mentioned in this book.

Typical mechanical properties		
Tensile strength (N/mm²)	Elongation (%)	Uses
280 510	55 4	*Gilding metal*—used for imitation jewellery, because of its gold-like colour, good ductility, and its ability to be brazed and enamelled.
320 700	70 5	*Cartridge brass*—deep-drawing brass of maximum ductility. Used particularly for the manufacture of cartridge- and shell-cases.
320 700	65 4	*Standard brass*—a good general-purpose cold-working alloy when the high ductility of 70–30 brass is not required. Widely used for press-work and limited deep-drawing.
340 725	55 4	*Basis brass*—a general-purpose alloy, suitable for limited cold-working operations.
370	40	*Yellow or Muntz metal*—hot-rolled plate. Can be cold-worked only to a limited extent. Also extruded as rods and tubes.
450	30	*Free-cutting brass*—very suitable for high-speed machining, but can be deformed only slightly by cold-work.
420	35	*'Naval' brass*—structural uses, also forgings. Tin raises corrosion-resistance, especially in sea water.
470 minimum 540 minimum	20 minimum 15 minimum	*High-tensile brass* ('manganese bronze')—pump-rods, stampings, and pressings. Also as marine castings such as propellors, water-turbine runners, rudders, etc.

16.51 Brasses with more than 37% zinc contain the phase β'. This is a hard, somewhat brittle substance; so a 60–40 brass lacks ductility. If such a brass is heated to 454°C (fig. 16.2), the phase β' changes to β, which is soft and malleable. As the temperature is increased further, the α phase present dissolves in β, until at X the structure is entirely malleable β. Therefore, 60–40 type brasses are best hot-worked at about 700°C. This treatment also breaks up the coarse cast structure, and replaces it with a fine granular structure.

Thus, brasses with less than 37% zinc are usually cold-working alloys, whilst those with more than 37% zinc are hot-working alloys. A copper-zinc alloy containing more than 50% zinc would be useless for engineering purposes, since it would contain the very brittle phase γ.

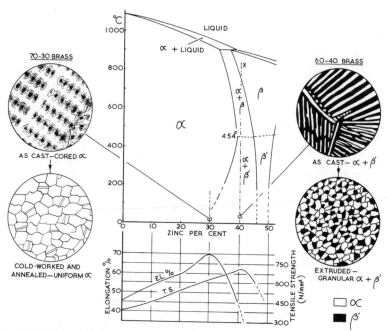

Fig. 16.2—The copper–zinc thermal equilibrium diagram, showing representative structures and properties of the two main classes of brass—70-30 and 60-40 alloys.

Up to 1% *tin* is sometimes added to brasses, to improve their resistance to corrosion, particularly under marine conditions. *Lead* is insoluble in both molten and solid brass, and exists in the structure as tiny globules. About 2% lead will improve the machinability (21.23) of brass.

16.52 *Manganese, iron*, and *aluminium* all increase the tensile strength of a brass, and are therefore used in high-tensile brasses. These alloys are sometimes known, rather misleadingly, as 'manganese bronzes'. Additions of up to 2% aluminium are also used to improve the corrosion-resistance of some brasses.

The more important brasses are shown in table 16.1.

Tin bronzes

16.60 Tin bronze was almost certainly the first metallurgical alloy to be used by Man, and it is a sobering thought that, for roughly a half of the 4000 years of history during which he has been using metallurgical alloys, bronze was his sole material. Though tin bronzes have relatively limited uses these days—owing mainly to the high prices of both copper and tin—they still find application for special purposes. Tin bronzes contain up to 18% tin, sometimes with smaller amounts of phosphorus, zinc, or lead.

16.61 The copper-tin equilibrium diagram in its entirety is rather complex, and its interpretation is made more difficult by the fact that bronzes need to be cooled very slowly indeed—far more slowly than is likely to be encountered industrially—for them to reach true equilibrium. In fact, the complete equilibrium diagram* shows that a structural change occurs at 350°C. Under normal industrial cooling rates, this change never occurs; so, in the diagram used here (fig. 16.3), the lower part has been omitted deliberately, as being irrelevant to our studies. All microstructures obtained in commercially produced bronzes can be explained by reference to the diagram given here.

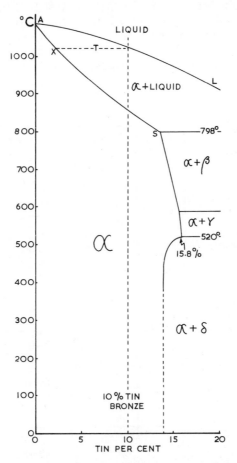

Fig. 16.3—The copper–tin thermal equilibrium diagram, somewhat simplified to suit industrial conditions of cooling.

* Readers who are curious can refer to *Engineering Metallurgy, Part 1* (fig. 16.3), by this author, for the complete diagram.

The diagram suggests that a cast bronze containing 10% tin might be expected to be completely α in structure, and therefore tough and ductile. However, the fact that the liquidus and solidus lines (*AL* and *AS*) are far apart means that coring of cast bronzes is likely to be excessive. Thus, as the 10% alloy begins to solidify (at temperature *T*), it will form dendrites of α of composition *X* (about 2% tin) to begin with. If the casting cools quickly, this coring will remain, and, since the *overall* composition of the alloy is 10% tin, it follows that the last portion of liquid to solidify may contain as much as 17% tin. At this composition, the structure will consist of α + δ, as indicated by the diagram. Now δ is a hard, extremely brittle compound (pale blue in colour); so a cast 10% tin bronze will be brittle, because of this δ phase distributed at the fringes of the α crystals.

If the alloy were annealed at about 700°C for seven or eight hours, diffusion would occur; that is, copper and tin atoms would move within the crystals, so that the structure would become more uniform. Thus δ

BS specification number	Composition %			Condition
	Cu	Sn	Other elements	
—	95·5	3	Zn—1·5	Soft Hard
2870/PB 101	96	3·75	P—0·1	Soft Hard
2870/PB 102	94	5·5	P—0·2	Soft Hard
1400/PB1—C	89	10	P—0·5	Sand cast
—	81	18	P—0·5	Sand cast
1400/G1—C	88	10	Zn—2	Sand cast
1400/LG2—C	85	5	Zn—5 Pb—5	Sand cast
1400/LB5—C	75	5	Pb—20	Cast

Table 16.2—Tin bronzes and phosphor bronzes.

would be absorbed; so the resultant structure would become one of uniform α, as indicated by the diagram. The material would then be tough and ductile. Due to coring effects associated with rapid cooling, bronzes with as little as 6% tin may contain small amounts of δ in the cast state, but are generally sufficiently ductile to permit their being cold-worked. During subsequent annealing processes, the small amount of δ is soon absorbed.

Thus, commercial tin bronzes can be divided into two groups.

(1) Wrought tin bronzes containing up to approximately 7% tin. These alloys are generally supplied as rolled sheet and strip, or as drawn rod, wire, and turbine-blading.

(2) Cast tin bronzes containing 10 to 18% tin, used mainly for high-duty bearings.

16.62 *Phosphor bronze.* Most of the tin bronzes mentioned above contain

Typical mechanical properties		Uses
Tensile strength (N/mm²)	Elongation (%)	
320 725	65 5	*Coinage bronze*—British 'copper' coinage now contains rather less tin (0·5%) and more zinc (2·5%).
340 740	65 15	*Low-tin bronze*—springs and instrument parts. Good elastic properties and corrosion-resistance.
350 700	65 15	*Drawn phosphor bronze*—generally used in the work-hardened condition; steam-turbine blading. Other components subjected to friction or corrosive conditions.
280	15	*Cast phosphor bronze*—supplied as cast sticks for turning small bearings, etc.
170	2	*High-tin bronze*—bearings subjected to heavy loads —bridge and turntable bearings.
290	16	*Admiralty gunmetal*—pumps, valves, and miscellaneous castings (mainly for marine work, because of its high corrosion-resistance); also for statuary, because of good casting properties.
220	13	*Leaded gunmetal* (or 'red brass')—a substitute for Admiralty gunmetal; also where pressure tightness is required.
155	6	*Leaded bronze*—a bearing alloy; can be bonded to steel shells for added strength.

up to 0·05% phosphorus, left over from the deoxidation process which is carried out before casting. Sometimes, however, phosphorus is added—in amounts up to 1·0%—as a deliberate alloying element, and only then should the material be termed 'phosphor bronze'. The effect of phosphorus is to increase the tensile strength and corrosion-resistance, whilst, in the case of cast bearing alloys, reducing the coefficient of friction.

16.63 *Gunmetal* contains 10% tin and 2% zinc, the latter acting as a deoxidiser, and also improving fluidity during casting. Since zinc is considerably cheaper than tin, the total cost of the alloy is reduced. Gunmetal is no longer used for naval armaments, but it is used as a bearing alloy, and also where a strong, corrosion-resistant casting is required.

16.64 *Coinage bronze* is a wrought alloy containing 3% tin and, in the interests of economy, 1·5% zinc. In Britain even more of the tin has been replaced by zinc (see Table 16.2).

16.65 *Leaded bronzes*. Up to 2·0% lead is sometimes added to bronzes, in order to improve machinability. Some special bearing bronzes contain up to 24% lead, and will carry greater loads than will 'white metal' bearings. Since the thermal conductivity of these bronzes is also high, they can work at higher speeds, as heat is dissipated more quickly.

Aluminium bronzes
16.70 Like brasses, the aluminium bronzes can be divided into two groups: the cold-working alloys, and the hot-working alloys. Those alloys con-

BS specification number	Composition %			Condition
	Cu	Al	Other elements	
2870/CA 101	95	5	—	Soft Hard
2870/CA 102	Remainder	7·5	Fe, Mn, and Ni up to 2·0 total	Hot-worked
2872/CA 104	80	10	Fe—5 Ni—5	Forged
1400/AB1—B	Remainder	9·5	Fe—2·5 Ni and Mn up to 1·0 each (optional)	Cast

Table 16.3—Aluminium bronzes.

taining approximately 5% aluminium are ductile and malleable, since they are completely solid solution in structure; consequently, they have a good capacity for cold-work. Since they also have a good resistance to corrosion, and a colour similar to that of 22 carat gold, they are widely used for decorative purposes, cheap jewellery, imitation wedding-rings, and the like.

The hot-working alloys contain in the region of 10% aluminium, and, if allowed to cool slowly, the structure is brittle, due to the precipitation of a hard compound within it. When this structure is heated to approximately 800°C, it changes to one which is completely solid solution, and hence malleable; so alloys of this composition can be hot-worked successfully. Similar alloys are also used for casting to shape by both sand- and die-casting methods. To prevent precipitation of the brittle compound mentioned above, castings are usually ejected from the mould as quickly as possible, so that they cool rapidly.

An interesting feature of the 10% alloy is that it can be heat-treated in a manner similar to that for steel. A hard martensitic type of structure is produced on quenching from 900°C, and its properties can be modified by tempering. Despite these apparently attractive possibilities, heat-treatment of aluminium bronze is not widely employed, and such of these alloys as are used find application mainly because of their good corrosion-resistance, retention of strength at high temperatures, and good wearing properties.

Aluminium bronze is a difficult alloy to cast successfully, because, at its casting temperature (above 1000°C), aluminium oxidises readily. This

Typical mechanical properties		Uses
Tensile strength (N/mm²)	Elongation (%)	
390 770	70 4	Decorative purposes—imitation jewellery; also some engineering applications, mainly in tube form. Good resistance to corrosion and to oxidation on heating.
430	45	Chemical engineering, particularly at fairly high temperatures.
725	20	Forged propellor-shafts, spindles, etc. for marine work. Can be heat-treated by quenching and tempering.
520	30	The most widely used aluminium bronze for both die- and sand-casting. Used in chemical plant and marine conditions—pump-casings, valve-parts, gears, propellors, etc.

leads to aluminium oxide dross becoming entrapped in the casting, unless special casting techniques are employed, and an increase in the cost of the process inevitably results. Compositions and uses of some aluminium bronzes are given in table 16.3.

Copper–nickel alloys

16.80 The metals copper and nickel 'mix' in all proportions in the solid state; that is, a copper-nickel alloy of any composition consists of only one phase—a uniform solid solution. For this reason, *all* copper–nickel alloys are relatively ductile and malleable, since there can never be any brittle phase present in the structure. In the cast state, a copper–nickel alloy may be cored (9.36), but this coring can never lead to the precipitation of a brittle phase. In other words, the metallurgy of these alloys is very simple —and not particularly interesting as a result.

Cupro-nickels may be either hot-worked or cold-worked, and are shaped

BS specification number	Composition %			Condition
	Cu	Ni	Other elements	
2870/CN 101	93	5	Fe—1·2 Mn—0·5	Soft Hard
2870/CN 104	80	20	Mn—0·25	Soft Hard
2870/CN 105	75	25	Mn—0·25	Soft Hard
2870/CN 106	70	30	Mn—0·4	Soft Hard
3072/3076/NA13	29	68	Fe—1·25 Mn—1·25	Soft Hard
3073/3076/NA18	29	66	Al—2·75 Fe—1·0 Mn—0·4	Soft Hard Heat-treated
2870/NS 106	60	18	Zn—22 Mn—0·4	—
2874/NS 111	60	12	Zn—26 Pb—2 Mn—0·25	—

Table 16.4—Cupro-nickels and nickel-silvers.

by rolling, forging, pressing, drawing, and spinning. Their corrosion-resistance is high, and only the high cost of both metals limits the wider use of these alloys.

16.81 *Nickel silvers* contain from 10 to 30% nickel, and 55 to 65% copper, the balance being zinc. Like the cupro-nickels, they are uniform solid solutions. Consequently, they are ductile, like the high-copper brasses, but have a 'near white' colour, making them very suitable for the manufacture of forks, spoons, and other tableware. When used for such purposes, these alloys are usually silver plated—the stamp 'EPNS' means 'electroplated nickel silver'. The machinability of these—as of all other copper alloys—can be improved by the addition of 2% lead. Such alloys are easy to engrave, and are also useful for the manufacture of Yale-type keys, where the presence of lead makes it much easier to cut the blank to shape.

Typical mechanical properties		Uses
Tensile strength (N/mm^2)	Elongation (%)	
260 460	50 5	Has slightly better mechanical properties and corrosion-resistance than pure copper.
340 540	45 5	Used for bullet-envelopes, because of its high ductility and corrosion-resistance.
350 600	45 5	Mainly for coinage—the current British 'silver' coinage.
350 650	45 5	Condenser- and cooler-tubes, where high corrosion-resistance is necessary.
560 720	45 20	*Monel metal*—good mechanical properties and excellent corrosion-resistance. Chemical engineering plant, etc.
680 760 1060	40 25 22	'*K*' *Monel*—a heat-treatable alloy. Used for motor-boat propellor-shafts.
—	—	*Nickel silver*—spoons, forks, etc.
—	—	*Leaded nickel silver*—Yale-type keys, etc.

Other copper-base alloys

16.90 A number of copper-base alloys contain small quantities of alloying elements which are generally added to increase the tensile strength.

16.91 *Copper-beryllium* (or beryllium bronze) contains approximately 1·75% beryllium and 0·2% cobalt. It is a heat-treatable alloy which can be precipitation-hardened in a manner similar to that of some of the aluminium alloys. A glance at the equilibrium diagram (fig. 16.4) will show why this is so. At room temperature, the slowly cooled structure will consist of the solid solution α (in this case, almost pure copper), along with

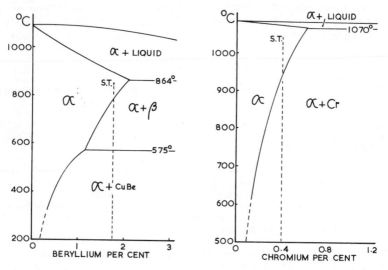

Fig. 16.4—Copper-base alloys which can be precipitation hardened.

In either case, the alloy is solution treated by heating it to the point ST, at which the structure becomes uniformly α. It is then quenched, to retain this structure, and then precipitation-hardened.

particles of the compound CuBe. If this is heated—that is, solution-treated—at about 800°C, the structure becomes completely α, as the CuBe is slowly dissolved by the solid solution. The alloy is then quenched, and in this condition is ductile, so that it can be cold-worked. If the alloy is now precipitation-hardened at 275°C for an hour, a tensile strength of up to 1400 N/mm^2 is obtained.

Since beryllium bronze is very hard in the cold-worked/heat-treated state, it is useful for the manufacture of non-sparking tools—chisels and hack-saw blades—for use in gas-works, 'dangerous' mines, explosives factories, or where inflammable vapours are encountered, as in paint and varnish works. Unfortunately, beryllium is a very scarce and expensive metal, and this limits its use in engineering materials.

16.92 *Copper-chromium* contains 0·5% chromium, and is also a heat-treatable alloy, as is indicated by the equilibrium diagram (fig. 16.4). It is used in some electrical industries, since it combines a high conductivity of about 80% with a reasonable strength of 550 N/mm^2.

16.93 *Copper-cadmium* contains about 1% cadmium, which raises the tensile strength of the hard-drawn alloy to 700 N/mm^2. Since the fall in electrical conductivity due to the cadmium is very small, this material is useful for telephone-wires, and other over-head lines.

16.94 *Copper-tellurium* contains 0·5% tellurium, which, being insoluble in solid copper, exists as small globules in the structure. Since tellurium is insoluble, it has little effect on electrical conductivity, but gives a big improvement in machinability (resembling lead in this respect).

16.95 *Arsenical copper* contains 0·4% arsenic, which increases from 200° to 550°C the temperature at which cold-worked copper begins to soften when it is heated. This type of copper was widely used in steam-locomotive fireboxes and boiler-tubes, and still finds use in high-temperature steam plant. It is useless for electrical purposes, however, because of the great reduction in electrical conductivity caused by the presence of arsenic in solid solution.

Chapter Seventeen
Aluminium and its Alloys

17.10 Minerals containing aluminium are difficult to decompose, and, for this reason, samples of the metal were not produced until in 1825 the Danish scientist H. C. Oersted used metallic potassium to chemically reduce aluminium from one of its compounds. Consequently, in those days aluminium cost about £250 per kg to produce, and was far more expensive than gold. It is reported that the more illustrious foreign visitors to the court of Napoleon III were privileged to use forks and spoons made from aluminium, whilst the French nobility had to be content with tableware in gold and silver. One still meets such cases of 'Jonesmanship', even in the metallurgical world—only today the writer was told of a presentation beer-tankard produced in the metal zirconium.

The extraction of aluminium
17.20 The modern electrolytic process for extracting aluminium was introduced simultaneously and independently in 1886 by Hall (in the USA) and Héroult (in France). Nevertheless, the metal remained little more than an expensive curiosity until the beginning of the present century. Since then, the demands by both air- and land-transport vehicles for a light, strong material have led to the development of aluminium technology and an increase in the production of the metal, until now it is second only to iron in terms of annual world production.

17.21 The only important ore of aluminium is bauxité, which contains aluminium oxide (Al_2O_3). Unfortunately, this cannot be reduced to the metal by heating it with coke (as in the case of iron ore), because aluminium atoms are, so to speak, too firmly combined with oxygen atoms to be detached by carbon. For this reason, an expensive electrolytic process must be used to decompose the bauxite and release aluminium. Since each kg of aluminium requires about 91 megajoules of electrical energy, smelting plant must be located near to sources of cheap hydro-electric power (HEP), often at great distances from the ore supply, and from the subsequent markets. Consequently, most aluminium is produced in the USA, and in Canada and Norway. HEP at Kinlochleven and Lochaber enables some aluminium to be smelted economically in Scotland, but 90% of the metal used in Britain is imported.

Crude pig iron can be purified (turned into steel) by blowing oxygen through it, to burn out the impurities (11.22), but this would not be possible in the case of aluminium, since the metal would burn away first, and leave us with the impurities. Instead, the crude bauxite ore is first purified by means of a chemical process, and the pure aluminium oxide is then decomposed by electrolysis. Since aluminium oxide has a very high melting-point, it is mixed with another aluminium mineral, cryolite, to form an electrolyte which will melt at a low temperature.

17.22 The furnace 'cell' (fig. 17.1) is usually about 2·5 m × 1·5 m × 0·6 m in size, and employs a current of 8000 to 30 000 A, at 7 V. The anodes, which gradually burn away, are made from a mixture of petroleum-coke and tar-pitch.

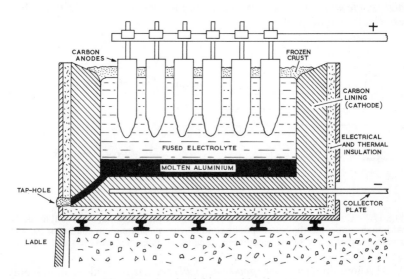

Fig. 17.1—An aluminium-smelting furnace.

When the electric current passes, aluminium particles, being positively charged, are attracted to the lining of the furnace, which constitutes the negative electrode (or cathode). Hence, molten aluminium collects at the bottom of the furnace, and is tapped off when necessary. In the meantime, oxygen is given off at the anodes, which burn as a result, and need to be replaced at frequent intervals.

Properties of aluminium
17.30 Although aluminium has a high affinity for oxygen, and might therefore be expected to oxidise (or 'rust') very easily, in practice it has *an excellent resistance to corrosion*. This is due largely to the thin but very dense film of oxide which forms on the surface of the metal and effectively protects it from further atmospheric attack. The reader will be familiar with the comparatively dull appearance of the surface of polished aluminium; this is due to the oxide film which immediately forms. The protective oxide skin can be artificially thickened by a process known as

'anodising'. Since aluminium oxide is extremely hard, anodising also makes the surface more wear-resistant.

17.31 The fact that it has a *high thermal conductivity* and good corrosion-resistance, and that any corrosion products which are formed are non-poisonous, makes aluminium very suitable for the manufacture of domestic cooking-utensils such as kettles, saucepans, and frying-pans. In the form of disposable collapsible tubes, it is used to contain a wide range of foodstuffs and toilet preparations, ranging from caviare to tooth-paste, whilst its *high malleability* makes it possible to produce very thin foil which is excellent for food-packaging.

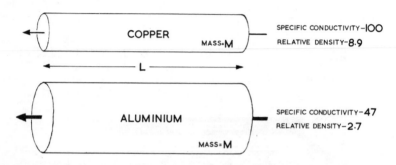

Fig. 17.2—Conductors of copper and aluminium of equal length, L, and equal mass, M. Although copper has the better specific conductivity, the aluminium conductor passes a greater current under similar conditions, because of its greater cross-section.

17.32 Aluminium has a *very good electrical conductivity*, which, though only about half that of copper, when considered weight for weight, can make aluminium a better proposition in some circumstances. Thus, if conductors of copper and aluminium of equal length and equal weight are taken, that of aluminium is a better overall conductor, because it has a greater cross-sectional area than that of the copper one. Aluminium conductors are used extensively in the grid system, generally strengthened by a steel core.

17.33 For use as a constructional material, pure aluminium lacks strength. In the 'soft' condition, its tensile strength is only 90 N/mm², whilst even in the work-hardened state it is no more than 135 N/mm². Hence, for most engineering purposes, aluminium is alloyed, in order to give a higher strength/weight ratio. Some of the high-strength alloys have a tensile strength in excess of 600 N/mm² when suitably heat-treated.

Alloys of aluminium
17.40 Many aluminium alloys are used in the wrought form; that is, they are rolled to sheet, strip, or plate; drawn to wire; or extruded as rods or

tubes. Other alloys are cast to shape by either a sand-casting or a die-casting process. In either case, some of the alloys may receive subsequent heat-treatment, in order further to improve their mechanical properties by inducing the phenomenon originally known as 'age-hardening', but now more properly termed 'precipitation-hardening'.

17.41 Thus the engineering alloys can be conveniently classified into four groups:

alloys which *are not* heat-treated	(1) wrought alloys	(2) cast alloys
alloys which *are* heat-treated	(3) wrought alloys	(4) cast alloys

In pre-war days, a bewildering assortment of aluminium alloys confronted the engineer. Worse still, they were covered by a rather untidy system of specification numbers, and each manufacturer used his own particular brand-name to describe an alloy. Since the Second World War, however, the number of useful alloys has been somewhat streamlined, and a systematic method of specification designation has been established by the British Standards Institution. In this system, each alloy is identified by a number; whilst prefix letters indicate its form, and whether or not it is a heat-treatable alloy. Suffix letters are used to show what treatment it has received.

17.42 *Wrought alloys* are covered by BS 1470/1475, and letters included in a specification number provided information as follows.

(*a*) Heat-treatment (first prefix letter):

   N—non-heat-treatable alloy
   H—heat-treatable alloy

(*b*) Form of material (second prefix letter). The more important forms are:

   S—plate, sheet and strip (BS 1470)
   T—drawn tube (BS 1471)
   F—forgings and forging stock (BS 1472)
   R—rivet, bolt and screw stock (BS 1473)
   E—bars, extruded round tube and sections (BS 1476)
   G—wire (BS 1475)

(*c*) Condition (suffix letter(s)):

   O—annealed to its lowest strength
   M—material 'as manufactured', e.g. rolled, drawn, extruded, etc.
   D—material solution-treated and then drawn
   H1, H2, H3, H4, Strain hardened—material subjected to cold work
   H5, H6, H7, H8. after annealing or to a combination of cold work and
                    partial annealing. Designations are in order of
                    ascending tensile strength.

TB—material solution-treated and aged naturally

TD—material solution-treated, cold worked and aged naturally

TE—material cooled from an elevated-temperature shaping process and precipitation treated

TF—material solution treated and precipitation treated

TH—material solution treated, cold-worked and then precipitation treated

Examples

(1) The specification number NS 5–H2, refers to the non-heat-treatable alloy no. 5, in the form of partly hardened sheet or strip.

(2) The number HT30–TF refers to the heat-treatable alloy no. 30, in tube form which has been solution-treated and then precipitation-hardened.

17.43 *Cast alloys* are covered by BS 1490, and specification numbers give information as to:

(*a*) Form of material:

Ingots—the alloy has *no* suffix letter

Castings—the suffix letter denotes the condition of the material

BS material number	Composition %				Condition
	Mn	Si	Mg	Others	
1470/NS 1(C) 1471/NT 1 (C) 1474/NE 1 (C)	99% aluminium 0·1 0·5 Maximum values		—	Fe—0·7	O H8
1470/NS 3 1475/NG 3	1·2	—	—	—	O H8
1470/NS 4 1471/NT 4 1472/NF 4 1474/NE 4 1475/NG 4	—	—	2·25	—	O H6
1470/NS 5 1471/NT 5 1472/NF 5 1473/NR 5 1474/NE 5 1475/NG 5	—	—	3·5	—	O H2
1473/NR 6 1475/NG 6	—	—	5·0	—	O H2

Table 17.1—Wrought aluminium alloys—not heat-treated.

(*b*) Condition:

M—'as cast' with no further treatment
TS—stress-relieved only
TE—precipitation treated
TB—solution treated
TB7—solution treated and stabilised
TF—solution treated and precipitation treated
TF7—full treatment, plus stabilisation

All of the alloy numbers representing cast alloys are prefixed by the letters LM. Thus LM 10 denotes alloy no. 10 in the ingot form, whilst LM 10-TF represents a casting of the same alloy in the fully heat-treated condition.

Wrought alloys which are not heat-treated
17.50 These are all materials in which the alloying elements form a solid solution in the aluminium, and this is a factor which contributes to their high ductility and good corrosion-resistance. Since these alloys are not heat-treatable, the necessary strength and rigidity can be obtained only

Typical mechanical properties		Uses
Tensile strength (N/mm²)	Elongation (%)	
90	35	Panelling and moulding, hollow-ware, electrical conductors, equipment for chemical, food- and brewing-plant, packaging
155	5	
105	34	Metal boxes, bottle-caps, food-containers, cooking-utensils, panelling of land-transport vehicles
200	4	
185	24	Marine super-structures, panelling exposed to marine atmospheres, chemical plant, panelling for road and rail vehicles, fencing-wire
250	5	
220	18	Ship-building, deep-pressing for car bodies
280	8	
260	18	Ship-building, and applications requiring high strength and corrosion-resistance, rivets
290	8	

by controlling the amount of cold-work in the final shaping process. They are supplied as 'soft' (O), 'quarter hard' ($\frac{1}{4}$H), 'half hard' ($\frac{1}{2}$H), 'three-quarter hard' ($\frac{3}{4}$H) or 'full hard' (H).

17.51 The commercial grades of aluminium—1(A), 1(B) and 1(C)—are sufficiently strong and rigid for some purposes, and the addition of up to 1·5% manganese (NS 3) will produce a slightly stronger alloy. The aluminium-magnesium alloys have a very good resistance to corrosion, and this corrosion-resistance increases with the magnesium content, making them particularly suitable for use in marine conditions (table 17.1).

Cast alloys which are not heat-treated

17.60 This group consists of alloys which are widely used for both sand-casting and die-casting. Rigidity, good corrosion-resistance, and fluidity during casting are their most useful properties.

17.61 The most important alloys in this group are those containing between 10 and 13% silicon. Alloys of this composition are approximately eutectic in structure (fig. 17.3). This makes them particularly useful for die-casting, since their freezing-range will be short, so that the casting will solidify quickly in the mould, making rapid ejection possible. In the ordinary cast condition (fig. 17.3) the eutectic structure is rather coarse, a factor which causes the alloy to be rather weak and brittle. However, the

BS 1490 designation	Composition %					Condition
	Si	Cu	Mg	Mn	Ni	
LM 2M	10	1·6	—	—	—	Chill cast
LM 4M	5	3	—	0·5	—	Sand cast Chill cast
LM 5M	—	—	4·5	0·5	—	Sand cast Chill cast
LM 6M	12	—	—	—	—	Unmodified Modified

Table 17.2—Cast aluminium alloys—*not* heat-treated.

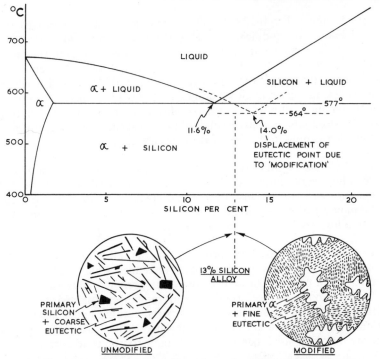

Fig. 17.3—The aluminium-silicon thermal equilibrium diagram.
The effects of 'modification' on both the position of the eutectic point and the structure are also shown.

Typical mechanical properties		Uses
Tensile strength (N/mm²)	*Elongation* (%)	
250	3	General purposes, particularly pressure die-castings Moderate strength alloy.
155 190	4 —	Sand-castings, gravity and pressure die-castings. Good foundry characteristics. An inexpensive general-purpose alloy, where mechanical properties are of secondary importance.
170 200	6 10	Sand-castings and gravity die-castings. Suitable for moderately-stressed parts. Good resistance to marine corrosion. Takes a good polish ('Birmabright').
125 200	5 15	Sand-, gravity- and pressure die-castings. Excellent foundry characteristics. Large castings for general and marine work. Radiators, sumps, gear-boxes, etc. One of the most widely used aluminium alloys.

structure—and hence the mechanical properties—can be improved by a process known as 'modification'. This involves adding a small amount of metallic sodium (only 0·01 % by weight of the total charge) to the molten alloy, just before casting. The resultant casting has a very fine grain, and the mechanical properties are improved as a result (table 17.2). The eutectic point is also displaced due to modification; so a 13 % silicon alloy, which would contain a little brittle primary silicon in the unmodified condition now contains tough primary α solid solution instead, since the composition of the alloy is now to the *left* of the *new* eutectic point.

17.62 High casting-fluidity and low shrinkage make the aluminium–silicon alloys very suitable for pressure die-casting. (Alloys of high shrinkage cannot be used, because they would crack on contracting in a rigid metal mould.) Since they are also very corrosion-resistant, the aluminium–silicon alloys are useful for marine work.

17.63 The most important property of the aluminium–magnesium–manganese alloys is their good corrosion-resistance, which enables them to receive a high polish. The well known 'Birmabright' alloys are noted for high corrosion-resistance, rigidity, and toughness, making them very suitable for use as moderately stressed parts working under marine conditions. Other casting alloys containing up to 10 % of either copper or zinc are now little used, because of their poor resistance to corrosion.

Wrought alloys which are heat-treated

17.70 In 1906, a German research metallurgist, Dr Alfred Wilm, was investigating the effects of quenching on the mechanical properties of some aluminium alloys containing small amounts of magnesium, silicon, and copper. To his surprise, he found that, if quenched test-pieces were allowed to remain at room temperature for a few days, without further heat-treatment, a considerable increase in strength and hardness of the material occurred. This phenomenon, subsequently called 'age-hardening', was unexplained at the time, since no apparent change in the microstructure was detected, and it is only in recent years that a reasonably satisfactory explanation of the causes of 'age-hardening' has been evolved. However, there are many instances in which metallurgical practice has been established long before it could be explained in terms of underlying theory—the heat-treatment of steel is an example—and Wilm's discovery was soon developed in the form of the alloy 'duralumin', which found application in the structural frames of the airships of Count von Zeppelin, which bombed England during the First World War.

17.71 As in the case of steels, the heat-treatment of an aluminium alloy is related to the equilibrium diagram; and, in this instance, the important part of the diagram (fig. 17.4) is the sloping boundary line *ABCD*, the significance of which was dealt with in 9.61. Here the slope of *ABCD*

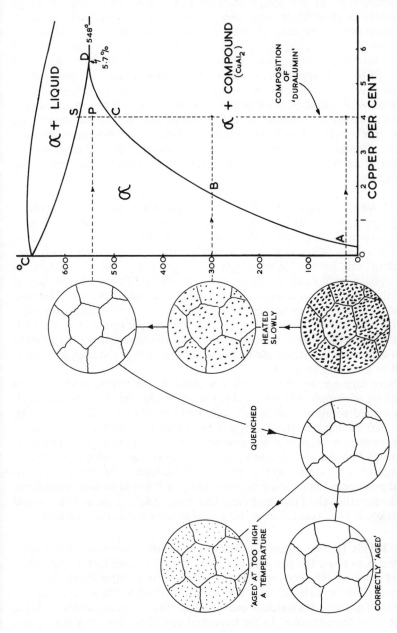

Fig. 17.4—Structural changes which take place during the heat-treatment of a duralumin-type of alloy.

indicates that, as the temperature rises, the amount of copper which dissolves in *solid* aluminium also increases. This is a fairly common phenomenon in the case of ordinary liquid solutions; for example, the amount of salt which will dissolve in water increases as the temperature of the solution increases. Similarly, if the hot, 'saturated' salt solution is allowed to cool, crystals of solid salt begin to separate—or 'precipitate', as we say—from the solution, so that 'equilibrium' is maintained. In the case of *solid* solutions, these processes of solution and precipitation take place more slowly, because atoms find greater difficulty in moving in solid solutions than they do in liquid solutions, where all particles can move more freely. Consequently, once a solid solution has been formed at some high temperature by heat-treatment, its structure can usually be trapped by quenching.

17.72 We will assume that we have an aluminium-copper alloy containing 4% copper, since this is the basic composition of some of the 'duralumin' alloys in use. Suppose the alloy has been extruded in the form of rod, and has been allowed to cool slowly to room temperature. This slow cooling will have allowed plenty of opportunity for the copper to precipitate from solid solution—not as particles of pure copper, but as small crystals of an aluminium-copper compound of chemical formula $CuAl_2$.* Since this compound is hard and brittle, it will render the alloy brittle. Moreover, since only about 0·2% of copper (point A in fig. 17.4) has remained behind in solution in the solid solution α, this solid solution will lack strength. Hence the mechanical properties of the alloy as a whole will be unsatisfactory.

Now suppose the alloy is slowly heated. As the temperature rises, the particles of compound $CuAl_2$ will gradually be absorbed into the surrounding α solid solution (just as salt would be dissolved by water). Thus at, say, 300°C, much of the $CuAl_2$ will have been dissolved, and an amount of copper equivalent to B will now be in solution in the α. The amount of copper which dissolves increases with the temperature, and solution will be complete at C, as the last tiny particles of $CuAl_2$ are absorbed. In industrial practice, a temperature corresponding to P would be used, in order to make sure that all of the compound had been dissolved, though care would be taken not to exceed S, as at this point the alloy would begin to melt.

17.73 This heating process is known as *solution-treatment*, because its object is to cause the particles of $CuAl_2$ to be dissolved by the aluminium solid solution (α). Having held the alloy at the solution-treatment temperature long enough for the $CuAl_2$ to be absorbed by the α solid solution, it is then quenched in water, in order to 'trap' this structure, and so preserve it at room temperature. In the quenched condition, the alloy is stronger,

---

* This formula indicates that two atoms of aluminium are combined with each atom of copper.

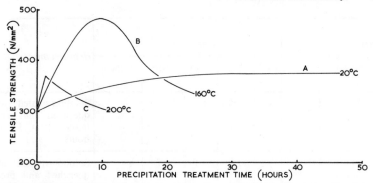

Fig. 17.5—The effects of time and temperature of precipitation-treatment on the strength of duralumin.

because now the whole of the 4 % copper is dissolved in the α; but it is also more ductile, because the brittle crystals of $CuAl_2$ are now absent. If the quenched alloy is allowed to remain at room temperature for a few days, its strength and hardness gradually increase (with a corresponding fall in ductility). This phenomenon, which was known as *age-hardening*, is due to the fact that the quenched alloy is no longer in 'equilibrium', and tries its best to revert to its original $\alpha + CuAl_2$ structure by attempting to precipitate particles of $CuAl_2$. In fact, the copper atoms do not succeed in moving very far within the aluminium structure, but the change is sufficient to cause a bigger resistance to the movement of slip planes within the alloy, i.e. the strength has increased.

17.74 This internal change within the alloy can be accelerated and made to proceed further by a 'tempering' process, formerly referred to as *'artificial' age-hardening*. A typical treatment may consist of heating the quenched alloy for several hours at, say, 160°C, when both strength and hardness will be found to have increased considerably (curve B in fig. 17.5). The term *'precipitation-hardening'* is now used to describe the increase in hardness produced, both by this type of treatment and by that which occurs at ordinary temperatures, as mentioned earlier. Care must be taken to avoid using too high a temperature during this precipitation treatment, or visible particles of $CuAl_2$ may form in the structure (fig. 17.4). If this happens, the mechanical properties will already have begun to fall again (curve C in fig. 17.5).

17.75 Because of the need for accurate temperature control during the heat-treatment of these alloys, solution-treatment is often carried out in salt baths; whilst precipitation-treatment generally takes place in air-circulating furnaces.

17.76 The aluminium–copper alloys are by no means the only ones which can be precipitation-hardened. Aluminium alloys containing small

Relevant BS and other specifications	Composition (%)					Heat-treatment
	Cu	Si	Mg	Mn	Others	
1474/HE9—TF	—	0·5	0·7	—	—	Solution-treated at 520 quenched and precipitati hardened at 170°C for hours.
1471/HT30—TF	—	1·0	1·0	0·7	—	Solution-treated at 510 quenched and precipitati hardened at 175°C for hours.
BS 2L84* BS 2L85	1·5	1·0	0·8	—	—	Solution-treated at 525 quenched and precipitati hardened at 170°C for hours.
BS 3L70* BS 3L77	4·1	0·5	0·8	0·7	Ti } 0·3 Cr } opt.	Solution-treated at 480 quenched and 'aged' at ro temperature for 4 days.
1470/HS15—TF	4·3	0·7	0·5	0·7	Ti } 0·3 Cr } opt.	Solution-treated at 510 quenched and precipitati hardened at 170°C for hours.
1471/HT20—TF	0·3	0·6	1·0	Either 0·5 Mn or 0·25 Cr		Solution-treated at 520 quenched and precipitati hardened at 170°C for hours.
BS 2L88* BS 2L95	1·6	0·5	2·5	0·3	Zn—7·0	Solution-treated at 465 quenched and precipitati hardened at 120°C for hours.

Table 17.3—Wrought aluminium alloys—heat-treated

*BS Aerospace Series, Section L (aluminium and the light alloys)

amounts of magnesium and silicon (forming Mg_2Si) can also be so treat as can numerous magnesium-base alloys (19.11), and also some copp base alloys (16.91 and 16.92). Compositions, treatment, and propert of some representative heat-treatable wrought aluminium alloys are giver table 17.3.

Cast alloys which are heat-treated

17.80 Some of these alloys are of the 4% copper type, as described abo but possibly the best known of them contains an additional 2% nickel a 1·5% magnesium. This is Y-alloy, 'Y' being the series letter used to ident

Typical mechanical properties		Uses
Tensile strength (N/mm²)	Elongation (%)	
250	10	Glazing bars and window sections; windscreen and sliding-roof sections for automobiles. Good corrosion-resistance and surface finish.
310	8	Structural members for road-, rail-, and sea-transport vehicles; ladders and scaffold tubes; overhead lines (high electrical conductivity); architectural work.
390	10	Structural members for aircraft and road vehicles; tubular furniture.
400	10	Stressed parts in aircraft and other structures; general purposes. The original 'duralumin'.
510	10	Highly stressed components in aircraft stressed-skin construction; engine parts, such as connecting rods.
310	13	Plates, bars, and sections for shipbuilding; body panels for cars and rail vehicles; containers.
650	11	Highly stressed aircraft structures, such as booms; other military equipment requiring high strength/weight ratio. The strongest aluminium alloy produced commercially.

t during its experimental development at the National Physical Laboratory, during the First World War. Whilst Germany was concentrating on the production of wrought duralumin for structural members of its Zeppelins; n Britain, research was being aimed at the production of a good heat-treatable *casting* alloy, for use in the engines of our fighter planes. The airframes of these wonderful machines were constructed largely of wood and 'doped' canvas, but a light alloy was needed from which high-duty pis-ons and cylinder-heads could be constructed for use at high temperatures. Y-alloy was the result of this research, and remains a popular material today. This and other heat-treatable casting alloys are enumerated in table 17.4.

BS specification number	Composition (%)					Heat-treatment
	Cu	Si	Mg	Mn	Others	
1490/LM4—TF	3	5	—	0·5	—	Solution-treated at 520°C for 6 hours; quenched in hot water or oil; precipitation-hardened at 170°C for 12 hours.
1490/LM9—TF	—	11·5	0·4	0·5	—	Solution-treated at 530°C for 2–4 hours; quenched in warm water; precipitation-hardened at 160°C for 16 hours.
1490/LM10—TF BS 4L53*	—	—	10	—	—	Solution-treated at 425°C for 8 hours; cooled to 390°C, and then quenched in oil at 160°C, or in boiling water.
BS 4L35*	4	—	1·5	—	Ni—2·0	Solution-treated at 510°C; precipitation-hardened in boiling water for 2 hours, or aged at room temperature for 5 days.
BS 3L51*	1·4	2·5	0·2	—	Fe—1·0 Ni—1·3 Ti—0·15	No solution-treatment required; precipitation-hardened at 165°C for 8–16 hours.
1490/LM29—TE	1	23	1	—	Ni—1.0	Chill cast and precipitation treated.

Table 17.4—Cast aluminium alloys—heat-treated.

* BS Aerospace Series, Section L (aluminium and the light alloys)

Typical mechanical properties		Uses
Tensile strength (N/mm²)	Elongation (%)	
320	1	General purposes (sand-, gravity-, and pressure die-casting).
290	2	Suitable for intricate castings, due to fluidity imparted by silicon. Good corrosion-resistance.
320	18	Good strength and ductility, coupled with good corrosion-resistance. Used in marine work (sea-planes).
280	—	Pistons and cylinder-heads for liquid- and air-cooled engines; general purposes. The original 'Y' alloy.
200	3	A good general-purpose alloy for sand-casting and gravity die-casting. High rigidity and moderate resistance to shock.
190	1	A gravity or pressure die-casting alloy—cylinder blocks and pistons in the automotive industries. A *hyper*-eutectic alloy in which the primary silicon is 'refined' by small additions of red phosphorus. Good wear resistance (Hardness—140 Brinell). Specification lays down microscopical examination to control crystal structure, i.e. size and distribution of primary silicon.

Chapter Eighteen
Nickel and its Alloys

18.10 Although the Ancient Chinese may have used alloys similar in composition to our nickel silvers (16.81), nickel itself was not discovered in Europe until about 1750. At that time, the copper smelters of Saxony were having trouble with some of the copper ores they were using, for, although these ores appeared to be of the normal type, they produced a metal most unlike copper. This metal was given the name of 'kupfer-nickel', which, somewhat liberally translated from the Old Saxon, meant 'copper possessed of the Devil'. Later the material was found to be an alloy containing a new metal which was allowed to retain the title of 'nickel' —or 'Old Nick's metal'.

18.11 The most important deposits of nickel ore are found in the region of Sudbury, Ontario. These ores are also worked for the copper, iron, sulphur, cobalt, gold, silver, and platinum they contain. The ore is first crushed, and then concentrated by 'flotation' processes, which, briefly, involve washing the crushed ore with water to which a suitable frothing-agent has been added. The waste 'gangue' sinks to the bottom of the tanks, but the mineral is not wetted, and is carried away in the froth. Nickel-rich, copper-rich, and iron-rich concentrates are separated from the ore in this way.

18.12 The nickel-rich concentrate is then smelted to produce a 'matte' (16.20) containing nickel sulphide and some copper sulphide. This matte is cooled slowly, so that the two sulphides separate to form distinct solid layers. The nickel sulphide may then be treated in a number of ways, in order to obtain the metal. In the Mond process, the nickel sulphide is first crushed, and then 'roasted' to form nickel oxide. This oxide is then finely ground, and reduced to the metal in a stream of 'water-gas'* at 350–400°C. The tiny particles of nickel so released then combine with carbon monoxide, present in the water-gas, to form a *volatile* compound called nickel carbonyl. Finally, this is passed over nickel pellets at a lower temperature, when the nickel carbonyl decomposes, depositing a film of nickel on the pellets.

Uses of commercially-pure nickel
18.20 Nickel is a 'white' metal, with a faintly greyish tint. Most of the uses of commercially-pure nickel depend upon the fact that it has a good resist-

---

* A mixture of hydrogen and carbon monoxide, produced by passing steam through white-hot coke.

ance to corrosion, not only by the atmosphere, but by many other reagents. Consequently, much nickel is used in the electroplating industries, not only as a finishing coat, but also as a 'foundation layer' for good quality chromium-plating.

18.21 Since nickel can be hot- and cold-worked successfully, and joined by most orthodox methods, it is used in the manufacture of chemical and food-processing plant; whilst nickel-clad steel ('Niclad') is used in the chemical and petroleum industries.

Nickel is used for the supporting wires and for the cathodes and grids in radio valves and in X-ray tubes,

Alloys of nickel
18.30 The greatest quantity of nickel is used in the manufacture of alloy steels (13.21, 13.23, 13.51); whilst other alloys containing considerable amounts of nickel include the cupro-nickels (16.80), the nickel silvers (16.81), and cast irons (15.62 and 15.81). Here we shall deal with those alloys in which nickel is the primary constituent, either in quantity or in terms of importance.

18.31 *Electrical resistance alloys for use at high temperatures.* These are generally nickel–chromium or nickel–chromium–iron alloys, the main features of which are:

(1) their ability to resist oxidation at high temperatures,
(2) high melting-ranges, and
(3) high electrical resistivity.

These properties make nickel–chromium alloys admirably suitable for the

Composition %			Resistivity $(10^{-8}\Omega m)$	Maximum working temperature (°C)	Uses
Ni	Cr	Fe			
80	20	—	103	1150	Heaters for electric furnaces, cookers, kettles, immersion-heaters, hair-dryers, toasters.
65	15	20	106	950	Similar to above, but for goods of lower quality; also for soldering-irons, tubular heaters, towel-rails, laundry-irons, and where operating temperatures are lower.
34	4	62	91	700	Cheaper-quality heaters working at low temperatures, but mainly as a resistance-wire for motor starter-resistances, etc.

Table 18.1—High-temperature resistance alloys.

Trade names	% Composition				Uses
	Ni	Mo	Fe	Others	
Corronel B	66	28	6	—	Resists attack by mineral acids and acid chloride solutions. Produced as tubes and other wrought sections for use in the chemical and petroleum industries, for constructing reaction-vessels, pumps, filter parts, valves, etc.
Ni-O-Nel	40	3	35	Cr-20 Cu-2	A 'Wiggin' alloy with characteristics similar to those of austenitic stainless steel, but more resistant to general attack, particularly in chloride solutions.
Hastelloy 'A'	58	20	22	—	Transporting and storing hydrochloric acid and phosphoric acid, and other non-oxidising acids.
Hastelloy 'D'	85	—	—	Si-10 Cu-3 Al-1	A casting alloy—strong, tough, and hard, but difficult to machine (finished by grinding). Resists corrosion by hot concentrated sulphuric acid.

Table 18.2—Corrosion-resistant nickel-base alloys.

manufacture of resistance-wires and heater-elements of many kinds, working at temperatures up to bright red heat. Representative alloys (the Henry Wiggin 'Brightray' series) are included in table 18.1.

18.32 *Corrosion-resistant alloys.* These alloys, for use at ordinary temperatures, all contain nickel, along with varying amounts of molybdenum and

Nimonic alloy	DTD specification	Approximate composition %					
		Cr	Ti	Al	Co	Mo	Ni
75	703B	20	0·5	—	—	—	BALANCE
80A	736A	20	2·2	1	—	—	
90	747A 5027	20	2·2	1·3	20	—	
115	5017	15	4	5	15	4	

Table 18.3—Some 'Nimonic' high-temperature alloys.

iron, and sometimes chromium and copper. Naturally, such alloys are relatively expensive, but their resistance to corrosion is extremely high. Consequently, they are used mainly in the chemical industries, to resist attack by strong mineral acids and acid chloride solutions—conditions which, in terms of corrosion, are about the most severe likely to be encountered industrially. A number of these alloys are described in table 18.2.

18.33 *High-temperature corrosion-resistant alloys.* A well known high-temperature alloy introduced many years ago is 'Inconel'. It contains 80% nickel, 14% chromium, and 6% iron, and is used for many purposes, including food-processing plant, hot-gas exhaust manifolds, and heating-elements for cookers. It is quite tough at high temperatures, because of the very low grain-growth imparted by nickel (13.21), and it does not oxidise appreciably, because of the protective film of chromium oxide which forms on the surface (13.50).

The properties of materials like Inconel were extended in the 'Nimonic' series of alloys introduced by Messrs Henry Wiggin—alloys which played a leading part in the development of the jet engine. These 'Nimonics' are basically nickel–chromium alloys which have been strengthened—or 'stiffened'—for use at high temperatures, by adding small amounts of titanium, aluminium, cobalt, and molybdenum in suitable combinations. These elements form phases within the structure, and have the effect of raising the limiting creep stress (4.51) at high temperatures. A few of the better-known Nimonic alloys are given in table 18.3.

18.34 *Low-expansion alloys.* Most materials expand when they are heated, and contract again as they cool. Some iron-nickel alloys, however, have extremely small coefficients of thermal expansion, making them very useful in numerous types of precision equipment operating under conditions of varying temperature.

Tensile strength (N/mm²) at:			Uses
°C	800°C	1000°C	
0	250	90	Gas-turbine flame tubes and furnace parts.
0	540	75	Gas-turbine stator-blades, after-burners, and other stressed parts working at high temperatures.
0	850	170	Rotor-blades in gas turbines.
0	1010	430	Excellent creep-resistant properties at high temperatures.

The coefficient of expansion is at a minimum for an iron-nickel alloy containing 36% nickel. This alloy, originally known as 'Invar', was often used for the pendulums of clocks requiring great accuracy, as in astronomical observatories. As the amount of nickel increases above 36%, the coefficient of expansion increases; so it is possible to produce alloys having a useful range of coefficients of expansion. Thus, alloys containing between 40 and 50% nickel have coefficients of expansion similar to those of many types of glass, so that efficient metal-glass seals can be produced in electric lamps, TV tubes, radio valves, and the like.

Many domestic and industrial thermostats depend for their operation on the uneven expansion of two different layers of metal in a bimetallic strip. One layer is usually of brass, which has a considerable coefficient of expansion; whilst the other is of one of these iron-nickel alloys, with a very low coefficient of expansion. The principle of operation of such a thermostat is indicated in fig. 18.1. Many readers will be familiar with the small thermostats of this type used in tropical aquaria. Some of these low-expansion alloys are detailed in table 18.4.

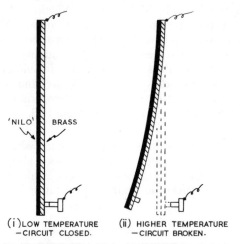

(i) LOW TEMPERATURE
—CIRCUIT CLOSED.

(ii) HIGHER TEMPERATURE
—CIRCUIT BROKEN.

Fig. 18.1—The principle of the bimetal thermostat.
As the temperature has increased, the brass layer has expanded more than the 'Nilo' layer, causing warping of the bimetal strip, and so breaking the contact.

Trade name	Composition %			Coefficient of expansion (at 20°C) × 10⁻⁶	Uses
	Ni	Fe	Co		
Nilo* 36 (Invar, Nivar)	36	64	—	0·9	Pendulum-rods, standard lengths, measuring-tapes, delicate precision sliding-mechanisms, thermostats for low-temperature operation.
Nilo 40	40	60	—	6·0	Thermostats for electric and gas cookers, heater-elements.
Nilo 42	42	58	—	6·2	Thermostats, also the core of copper-clad wire for glass seals in electric lamps, radio valves and TV tubes.
Nilo 50	50	50	—	9·7	Thermostats, also for sealing with soft glasses used in radio and electronic equipment.
Nilo K	29	54	17	5·7	Glass/metal seals in medium-hard glasses used in X-ray tubes and various electronic equipment.

* The trade name for these alloys, coined by Messrs Henry Wiggin.

Table 18.4—Iron-nickel low-expansion alloys.

Chapter Nineteen
Other Non-ferrous Metals and Alloys

Magnesium-base alloys

19.10 Magnesium is a fairly common metal in the Earth's crust and, of the metallurgically useful metals, only iron and aluminium occur more abundantly. However, magnesium is relatively expensive to produce, because it is difficult to extract from its mineral ores, electrolysis being used in a process similar to that employed for extracting aluminium. Because of its great affinity for oxygen, magnesium burns with an intensely hot flame, and was used as the main constituent of incendiary bombs during the Second World War; for this reason, it is also more difficult to deal with in the foundry than are other light alloys. To prevent its taking fire, it is melted under a layer of molten flux, and, during the casting process, 'flowers of sulphur' are shaken on to the stream of molten metal as it leaves the crucible. The sulphur burns in preference to magnesium. It is generally easy to identify a foundry-worker who is employed in casting magnesium—his eyebrows are always bleached by the sulphur dioxide gas evolved.

19.11 Magnesium is the lightest of the metals used in general engineering, having a relative density of only 1·7. Although pure magnesium is a relatively weak metal (tensile strength approximately 180 N/mm² in the annealed state), alloys containing suitable amounts of aluminium, zinc, and thorium can be considerably strengthened by precipitation-hardening treatments. Consequently, in view of these properties, magnesium-base alloys are used where weight is a limiting factor. Castings and forgings in the aircraft industry account for much of the magnesium alloys produced, e.g. landing-wheels, petrol-tanks, oil-tanks, crank-cases, and air-screws, as well as other engine parts in both piston and jet engines. The automobile industry too makes use of magnesium-base alloys, as, for example, in the ubiquitous Volkswagen.

19.12 In addition to the alloying elements mentioned above, small amounts of manganese, zirconium, and 'rare earth' metals are added to some magnesium alloys. Manganese helps to improve corrosion-resistance; whilst zirconium acts as a grain-refiner. The 'rare earths' (along with thorium) give a further increase in strength, particularly at high temperatures. These alloys, which are manufactured under such trade names as 'Magnuminium' and 'Elektron', include both cast and wrought materials, examples of which are given in tables 19.1 and 19.2.

Composition %						Condition	Tensile strength (N/mm²)
Al	Mn	Zn	Zr	Th	Rare earths		
8·0	0·3	0·7	—	—	—	As cast Precipitation-hardened	140 200
—	—	4·5	0·7	—	—	As cast Precipitation-hardened	230 260
—	—	4·2	0·7	—	1·2	As cast Precipitation-hardened	170 215
—	—	—	0·7	3·2	—	Precipitation-hardened	215

Table 19.1—Cast magnesium-base alloys.

Composition %					Condition	Tensile strength (N/mm²)
Al	Mn	Zn	Zr	Th		
—	1·5	—	—	—	Rolled Extruded	200 230
6·0	0·3	1·0	—	—	Forged Extruded	280 215
—	—	3·0	0·7	—	Rolled Extruded	260 310
—	1·0	—	—	3·0	Rolled	280

Table 19.2—Wrought magnesium-base alloys.

Zinc-base die-casting alloys

19.20 The growth of the die-casting industry was helped to a great extent by the development of modern zinc-base alloys. These are rigid and reasonably strong materials which have the advantage of a low melting-point, making it possible to cast them into relatively inexpensive dies. (Alloys of high melting-point will generally require dies in rather expensive heat-resisting alloy steels (table 13.6).)

19.21 A very wide range of components, both for the engineering industries and for domestic appliances, is produced in a number of different zinc-base alloys sold under the trade name of 'Mazak'. Automobile fittings such as door-handles and windscreen-wiper bodies account for possibly the largest consumption, but large quantities of these alloys are also used in electrical equipment, washing-machines, radios, alarm-clocks, and other domestic equipment. Compositions of the alloys most often used are given in table 19.3.

Composition %			Shrinkage (mm) after 5 weeks normal 'ageing'
Aluminium	Copper	Zinc	
4·0	2·7	Balance	0·001 12
4·0	1·0	,,	0·000 83
4·0	—	,,	0·000 65

Table 19.3—Zinc-base die-casting alloys.

19.22 During the development of these alloys, difficulty was experienced owing to the swelling of the casting during subsequent use, accompanied by a gradual increase in brittleness. These faults were found to be due to intercrystalline corrosion, caused by the presence of small quantities of impurities such as cadmium, tin, and lead. Consequently, very high-grade zinc of 'four nines' quality (i.e. 99·99% pure) is used for the production of these alloys.

19.23 Good quality zinc-base alloy castings undergo a slight shrinkage, which is normally complete in about five weeks (table 19.3). When close tolerances are necessary, a 'stabilising' anneal at 150°C for about three hours should be given before machining. This speeds up any volume change which is likely to occur.

Lead–antimony alloys
19.30 Antimony is a hard, extremely brittle metal, which consequently has only limited use in engineering alloys. It is added to bearing metals

Composition %		Uses
Antimony	Lead	
1	Balance	Cable sheathing
up to 4	,,	Collapsible tubes (artists' colours)
7	,,	Sheet used in chemical plant
7·5	,,	Accumulator plates
11	,,	Chemical-plant pipe-fittings

Table 19.4—Lead-antimony alloys.

(19.61), in order to reduce the coefficient of friction by increasing the hardness; whilst in the lead-base alloys, shown in table 19.4, it is added to increase both strength and hardness. Thus, a lead-base alloy containing more than 8 % antimony may be machined and screw-threaded. The main feature of these alloys is good resistance to corrosion, coupled with ease of forming.

Type-metals
19.40 Alloys containing lead, tin, and antimony are used for the manufacture of printing-type. Tin raises the toughness, and also improves the casting properties by increasing the fluidity and reducing the melting-point. Antimony increases both hardness and wear-resistance, and has the important effect of reducing contraction during solidification, thus ensuring that the alloy takes a good impression of the mould. Whilst most metals contract during solidification, antimony expands; hence, by adjusting the composition, an alloy of zero shrinkage can be produced.

Composition %			Uses
Antimony	Tin	Lead	
13–30	3–10	Balance	Type-metal
12	5	Balance	'Linotype' metal
12–23	5–17	Balance	'Stereotype' metal

Table 19.5—Type-metals.

Fusible alloys
19.50 These are generally eutectic mixtures containing three or four metals (9.70), usually of the group tin, lead, bismuth, antimony, cadmium, and mercury. Some of these alloys will melt well below the boiling-point of water, and, in less sophisticated days, were used for the manufacture of tea-spoons for practical jokers. Less frivolous uses in our serious modern world include plugs for automatic fire-extinguishing sprinklers, metal

Type of alloy	Composition %				Melting-point (°C)
	Bismuth	Lead	Tin	Others	
'Cerromatrix'	48	28·5	14·5	Antimony—9	102–225
Rose's alloy	50	28	22	—	100
Wood's alloy	50	24	14	Cadmium—12	71
Dental alloy	53·5	17·5	19	Mercury—10·5	60

Table 19.6—Fusible alloys.

patternwork, dental work, and the production of a temporary filling which will prevent collapse of the walls during the bending of a pipe. Like antimony, bismuth is a metal which expands on solidification, and alloys of the 'Cerromatrix' type are useful for setting press-tools in their holders, because of their slight expansion.

Bearing metals
19.60
The most important properties of a bearing metal are that it should be hard and wear-resistant, and have a low coefficient of friction. At the same time, however, it must be tough, shock-resistant, and sufficiently ductile to allow for 'running in' processes made necessary by slight misalignments. Such a contrasting set of properties is almost impossible to obtain in any single metallic phase. Thus, whilst pure metals and solid solutions are soft, tough, and ductile, they invariably have a high coefficient of friction, and consequently a poor resistance to wear. Conversely, intermetallic compounds are hard and wear-resistant, but are also brittle; so they have a negligible resistance to mechanical shock. For these reasons, bearing metals are generally compounded so as to give a suitable blend of phases, and generally contain small particles of a hard compound embedded in the tough, ductile background of a solid solution. During service, the latter tends to wear away slightly, thus providing channels through which lubricants can flow; whilst the particles of intermetallic compound are left standing 'proud', so that the load is carried with a minimum of frictional losses.

19.61 *White bearing metals* are either tin-base or lead-base. The former, which represent the better quality high-duty white metals, are known as 'Babbitt' metals, after Isaac Babbitt, their originator. All white bearing metals contain between 3·5 and 15% antimony, and much of this combines chemically with some of the tin, giving rise to an intermetallic compound. This forms cubic crystals ('cuboids'), which are easily identified in the microstructure (fig. 19.1). These cuboids are hard, and have low-friction properties; consequently, they constitute the necessary bearing surface in white metals. The background (or 'matrix') of the alloy is a tough, duc-

Type	BS specification	Composition (%)					Characteristics
		Sb	Sn	Cu	As	Pb	
Tin-base Babbitt metals	3332/1	7	90	3	—	—	These are generally heavy-duty bearing metals.
	3332/3	10	81	5	—	4	
	3332/6	10	60	3	—	27	
Lead-base bearing alloys	3332/7	13	12	0·75	0·2	Bal.	These alloys are generally lower-strength, lower-duty materials.
	3332/8	15	5	0·5	0·3	Bal.	
	—	10	—	—	0·15	Bal.	

Table 19.7—White bearing metals.

CUBOIDS OF
ANTIMONY–TIN COMPOUND

NEEDLES OF
COPPER–TIN COMPOUND

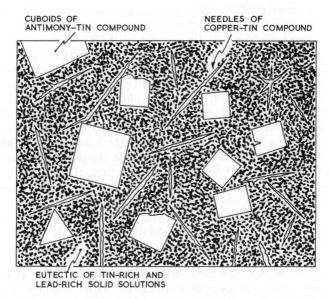

EUTECTIC OF TIN-RICH AND
LEAD-RICH SOLID SOLUTIONS

Fig. 19.1—A sketch of the structure of a typical white bearing metal. (×50).
The actual bearing surface is provided by the very hard cubic crystals of antimony–tin compound. These are embedded in a soft, tough eutectic, which will resist shock. The chief function of the copper is to form a network of needle-shaped crystals of copper–tin compound. These form first, as the alloy solidifies, and so prevent the antimony–tin cuboids (which form next) from floating to the surface of the cast bearing.

tile solid solution, consisting of tin with a little antimony dissolved in it.

In the interests of economy, some of the tin is generally replaced by lead. This forms a eutectic structure with the tin–antimony solid solution. The lead-rich white metals are intended for lower duty, since they can withstand only limited pressures. Some white bearing metals are detailed in table 19.7.

19.62 *Cadmium-base and zinc-base white metals* are employed as bearings to a limited extent. They are generally used only for light work, since, in respect both of ductility and anti-friction properties, they are inferior to the tin-base alloys.

19.63 *Copper-base bearing metals* include plain tin bronzes (10–15% tin) and phosphor bronzes (10–13% tin, 0·3–1·0% phosphorus). Both of these alloys follow the above-mentioned structural pattern of bearing metals. Some of the tin and copper combine to form particles of a very hard compound; whilst the remainder of the tin dissolves in the copper, to form a tough solid-solution matrix. These alloys are very widely used for bearings when heavy loads are to be carried.

19.64 For many small bearings in standard sizes, sintered bronzes are often used. These are usually of the self-lubricating type, and are made by mixing copper powder and tin powder in the proportions of a 90–10 bronze. Sometimes some graphite is added. The mixture is then 'compacted' at high pressure in a suitably shaped die, and is then sintered at a temperature which causes the tin to melt and so alloy with the copper, forming a continuous structure—but without wholesale melting of the copper taking place. The sintered bronze retains its porosity, and this is made use of in storing the lubricant. The bearing is immersed in lubricating oil, which is then 'depressurised' by vacuum treatment, so that, when the vacuum seal is broken, oil will be forced into the pores. In many cases, sufficient oil is absorbed to last for the life-time of the machine. Self-lubricating sintered bearings are used widely in the automobile industries and in other applications where long service with a minimum of maintenance is required. Consequently, many are used in domestic equipment such as vacuum cleaners, washing-machines, electric clocks, and gramophone-motors.

19.65 Leaded-bronzes are used in the manufacture of main bearings in aero-engines, and for automobile and diesel crankshaft bearings. They have a high wear-resistance and a good thermal conductivity, which helps in cooling them during operation. Brasses are sometimes used as low-cost bearing materials. They are generally of a low-quality 60–40 type, containing up to 1.0% each of aluminium, iron, and manganese.

19.66 *Plastic bearing materials* are also used, particularly where oil lubrication is impossible or undesirable. The best known substances are nylon (23.64 and table 23.1) and polytetrafluoroethylene (23.63 and table 23.1), both of which have low coefficients of friction. Polytetrafluoroethylene (PTFE, or 'Teflon') is very good in this respect, and in fact feels greasy to the touch. It is also used to impregnate some sintered-bronze bearings.

Chapter Twenty
Some Metals of the Space Age

20.10 One finds some difficulty in choosing a suitable title for this chapter. The description 'new metals' would not be altogether appropriate, since most of the metals in question have been known to chemists for many years, and only their applications are in any sense 'new'. Even the present title is inadequate, since it may suggest that these metals will ultimately replace those with which we are now familiar in general engineering practice. The truth is that steel is, and is likely to remain—in the foreseeable future at least—Man's most important metallurgical engineering material.

20.11 None the less, these 'space-age' metals have their particular uses, and in many cases they would be far more widely employed, were it not for their scarcity, or, in some instances, their high costs of extraction or fabrication. The more important of these metals, together with some of their physical properties, are included in table 20.1.

Beryllium
20.20 Beryllium was first discovered as long ago as 1797, but it was not until 1916 that small-scale commercial production of the metal began, in the USA. Since it is much lighter than aluminium, but at the same time has a higher melting-point, beryllium is a metal in which the modern aircraft industry has, predictably, shown great interest. Unfortunately, beryllium is a very scarce metal, and its only important ore is the semiprecious stone beryl. It also occurs in the precious stone emerald, but this can hardly be classed as an ore of the metal.

20.21 The extraction of beryllium from its ore is a difficult and expensive process. Unfortunately, the metallurgist's worries do not end even there, since beryllium lacks ductility, and proves to be a very difficult metal to work to shape. Moreover, it is an extremely poisonous substance, and great care is necessary to protect the workers from inhaling either its dust or its vapour. All of these adverse factors naturally add to the very high cost of producing any component from beryllium, and until recent years the only substantial use of the metal was as an addition to copper (16.91).

20.22 As with other 'space-age' metals, the technology of beryllium is being developed, with extruded and rolled sections (strip, wire, rods, and tubes) being produced, though the metal is given its primary shape most conveniently by means of powder-metallurgy methods (7.40). Beryllium is

Metal	Symbol	Relative density	Melting-point (°C)	Tensile strength (N/mm²)		Appearance
				Soft	Cold-worked	
Beryllium	Be	1·84	1277	300	675	Light, steely colour, fairly crystalline
Germanium	Ge	5·32	937	—	—	Lustrous, greyish-white, very brittle and crystalline
Hafnium	Hf	13·09	2222	450	720	A brilliant silver-grey metal
Niobium (Columbium)	Nb (Cb)	8·57	2468	270	555	A ductile, steely-grey metal
Plutonium	Pu	19·84	640	270 Cast	—	A white metal which tarnishes to a yellow colour
Tantalum	Ta	16·6	2996	200	410	A heavy, steely-blue metal, white when newly polished
Titanium	Ti	4·5	1668	230	400	A white metal with a 'fracture' like steel
Uranium	U	19·07	1132	370	550	Like polished iron in appearance, but tarnishes slowly
Zirconium	Zr	6·5	1852	210	600	A silvery-white metal; ductile and malleable when pure, but brittle when impure

Table 20.1—Some metals of the space age.

somewhat brittle at ordinary temperatures, but is stronger at elevated temperatures. Since it is so much lighter than aluminium, but has a higher melting-point than that metal, beryllium has found use in high-speed aircraft, and in rocketry.

Beryllium has properties which make it of interest to the nuclear-engineering industries, and much of the metal produced is used in the form of moderators and reflectors in nuclear reactors. It is easily penetrated by X-rays, and is therefore used as X-ray 'windows'. Other uses include parts for gyroscopes and computers.

Germanium

20.30 The naturally occurring elements can be divided into two groups: metals and non-metals. A few elements possess some properties which are metallic and some which are non-metallic, making it difficult to decide whether to classify them as metals or non-metals. They are often known as 'metalloids', and germanium is one of these.

20.31 The origins of the discovery of germanium make a very interesting story. In the 1860's, the Russian chemist Dimitri Mendeléef, classifying the then-known elements into groups, and comparing their properties, forecast the existence of another element with properties similar to those of silicon. He called the unknown element 'eka-silicon'. This proved a challenge to chemists of the day, and, when germanium was discovered a few years later, it was found to have properties very like those which Mendeléef forecast in his eka-silicon.

20.32 Like other metalloids, germanium has some properties which are metallic and some which are non-metallic. Thus it has a silvery metallic lustre, but is very brittle, so that it cannot be cold-worked. However, it has important electrical properties, which have resulted in its extensive use in transistors though more recently it has been largely replaced by silicon.

Zirconium

20.40 Zirconium is the tenth most abundant element in the earth's crust. Unfortunately, its ores never occur in highly concentrated deposits, as do those of many other metals, which, taking the earth's crust as a whole, may nevertheless be scarcer 'on average'. Moreover, zirconium is a very reactive metal; consequently, its compounds are difficult to decompose. Although discovered as long ago as 1789, little industrial use was made of zirconium until it was added to magnesium-base alloys (19.12), principally as a grain-refiner. The development of zirconium technology began in 1944, with the investigation of the properties of the high-purity metal.

20.41 Pure zirconium is a relatively soft, malleable, and ductile metal, which can be forged, hot- and cold-rolled, and drawn to wire. It is supplied as sheet, strip, foil, rod, tube, and wire. In the cold-worked state it is mechanically strong.

20.42 Zirconium is a chemically-reactive metal. In the form of wire, it is employed in some photographic flash-bulbs; whilst, alloyed with lead, it is used in lighter-flints. Nevertheless, it has a very good resistance to corrosion by the atmosphere and by many other reagents, including some acids. This is due to the formation of a thin but tenacious oxide film, which forms on the surface of the metal, and protects it from further corrosion. It resembles aluminium in this respect, and, like aluminium, it can be anodised

(22.441) to give added protection. Because of its high corrosion-resistance, zirconium finds use in the chemical industries, as agitators, pump and valve parts, heat-exchangers, etc.

20.43 By far the greatest use of zirconium is in the field of nuclear power. Its most important property is that it has a very low neutron-absorption, making it useful as a material for fuel containers and core components in water-cooled reactors. It was used in this way in the US submarine *Nautilus*.

A zirconium-base alloy, Zircalloy II, contains 1·5% tin, 0·12% iron, 0·1% chromium, and 0·05% nickel, and has a tensile strength of 510 N/mm² in the annealed state. This material is used as a fuel-canning material in the *Polaris* submarines, because of its low neutron-absorption.

Hafnium
20.50 Hafnium is certainly a metal of the twentieth century, since its existence was not finally verified until 1923. One reason for its late discovery was the fact that, chemically, hafnium is very similar to zirconium, and is in fact generally found in the ores of zirconium, where it had lurked unnoticed long after zirconium was discovered in 1789.

20.51 Like zirconium, hafnium has an excellent resistance to corrosion, and quite good mechanical strength. Although harder and less malleable than zirconium, it can be forged, rolled, and drawn, and consequently produced in most of the common forms, viz. sheet, strip, wire, rod, and tube.

20.52 Its properties differ from those of zirconium in one important field: whereas zirconium allows neutrons to pass through it quite freely, hafnium *absorbs* neutrons very effectively. Consequently, zirconium which is to be used as a nuclear fuel-canning material must be free of hafnium, and the difficult purification processes involved are very expensive. Since hafnium absorbs neutrons effectively, its main use is as a structural material and 'controller' in water-cooled nuclear reactors. It has been used in this way in nuclear-powered submarines. As a controller, it is employed in the form of rods, which can be passed into the reactor, and so control the energy level by 'blotting up' excess neutrons, thus reducing the rate at which fission occurs (20.92).

20.53 Hafnium has limited uses in non-nuclear fields, mainly in some rectifiers, and as lamp filaments and cathodes. Its wider use is limited by the high cost of separating it from zirconium.

Niobium
20.60 This metal was discovered near Connecticut, in 1801, by a British chemist who named it 'columbium', in honour of America. At about

the same time, a 'new' metal was isolated in Sweden. This proved to be an alloy of tantalum, and, when the new metal was separated from the alloy, in 1844, it was called 'niobium'—in Greek mythology, Niobe was the daughter of Tantalus. Soon after this, 'columbium' and 'niobium' were identified as being one and the same metal, but, since by then both names had become established, they continued to be used on their respective sides of the Atlantic. Even now, though the name niobium has been adopted by international agreement, 'columbium' is still sometimes used. Most niobium is found associated with tantalum ores, but, since these ores are scarce, niobium is an expensive metal to produce, and this limits its industrial use.

20.61 Niobium is quite malleable and ductile, and can be rolled to sheet and bar. It can be welded successfully, provided an adequate inert-gas shield (26.44) is used. Although it resists corrosion by most chemical substances, it is no better than tantalum in this respect, and, since the latter is the more plentiful, it is the more widely used.

Niobium has a very high melting-point (2468°C), and excellent strength at high temperatures. Since it also withstands oxidation at high temperatures, it is potentially useful for gas turbines and rocket motors, but scarcity at present limits its use.

20.62 It has been used as a fuel-canning material in high-temperature nuclear reactors (as at Dounreay), since it withstands attack by molten metals like sodium, which are used as coolants. However, it has little advantage here over the more readily available tantalum.

20.63 The bulk of niobium produced is used as an addition to austenitic stainless steels (13.53) and other alloys. Niobium–molybdenum alloys are used in rocket power-units, but the general use of niobium, either unalloyed or as a major alloying constituent, is limited by its scarcity and high cost.

Tantalum

20.70 The metal tantalum was produced in a reasonably pure form as long ago as 1820, though its industrial production did not begin until 1903, when the fact that it had a very high melting-point (2996°C) led to its use as a filament material for electric lamps. Its use increased in this direction until 1910, when it was replaced by tungsten, which had become much cheaper as its methods of production were developed.

20.71 Tantalum is a heavy metal with a steely-blue colour. Like aluminium, it protects itself with a thin but impermeable film of oxide; consequently, it has an excellent resistance to corrosion, and this fact influences most of its uses. It is superior even to platinum (which is *very* expensive) in respect of its resistance to corrosion by most chemical reagents; in fact, the corrosion-resistance of tantalum is roughly equal to that of glass.

Although tantalum has a very high melting-point, and is classed as a 'refractory metal', unlike most other refractory metals it is very ductile, so it can be rolled and drawn to very fine wire. At the same time, it is strong at high temperatures.

20.72 Since the Second World War, production of tantalum has increased considerably. It is used in stills, agitators, centrifugal pumps, pipes, and containers in the chemical trades. It is also useful as a plugging material, to repair holes in vitreous-lined tanks in the chemical and distilling industries. Spinnerets, used in the manufacture of artificial silk, are generally of tantalum. The cellulose or rayon threads issue through holes in the spinneret into an acid-bath which causes them to harden.

20.73 Tantalum has many uses in surgery, because, being very corrosion-resistant, it is non-toxic too, and does not irritate human tissue, which will grow on to it quite freely. Thus, it is used as cover wire for joining fractured bones, and as plates for replacing lost skull material. Since tantalum is very ductile, it can be drawn to fine wire, which is then woven to gauze for strengthening abdomen walls following hernia operations.

Though pure tantalum is a soft, ductile metal, its carbide is very hard, and is used in cemented-carbide cutting-tools.

Titanium
20.80 Titanium is probably the most important and most widely used of all these 'space-age metals'. As well as being strong and corrosion-resistant, it is also a very 'light' metal, and, since it has only just over half of the relative density of steel, this gives it an excellent strength/weight ratio. It is as corrosion-resistant as 18–8 stainless steel, but will also withstand the extreme corrosiveness of salt water.

20.81 Although titanium was discovered in Cornwall as long ago as 1791, by an English priest, W. Gregor, its industrial production did not begin until the 1940's. This was due partly to a lack of interest in the metal, but mainly because molten titanium reacts chemically with most other substances, making it difficult to extract, melt, and cast. Thus the modern Kroll process for titanium production operates entirely in vacuum, so that, although titanium is a fairly common element in combined form in the earth's crust,* it is likely to remain expensive, because of the difficulty of extracting, casting, and shaping it.

Titanium is a white metal with a fracture surface like that of steel. In the pure state, it has a maximum tensile strength of no more than 400 N/mm², but, when alloyed with small amounts of other metals such as aluminium, tin, and molybdenum, strengths of 1400 N/mm² or more can be obtained,

---

* It is some fifty times more plentiful than copper.

and, most important, maintained at much higher temperatures than is possible with aluminium alloys.

20.82 In view of the properties mentioned above—low relative density, high strength/weight ratio, good corrosion-resistance, and good strength at high temperatures—it is obvious that titanium will be used in increasing amounts in both aircraft and spacecraft. Titanium alloys have been used for some time for parts in jet engines, and in some modern aircraft about a quarter of the weight of the engine is taken up by titanium alloys. In structural members of aircraft, too, these alloys are finding increasing use, because of their high strength/weight ratio. For example, the use of titanium alloys to replace high-tensile steels in the Sikorsky helicopter led to a reduction in total weight of about 500 kg. The four engines of Concorde contain some 16 tonnes of titanium alloys, whilst they are widely used in military rocketry. Here an added advantage is the excellent resistance to corrosion of these alloys, since long periods of storage are essential to the strategy of the nuclear deterrent.

Uranium

20.90 In 1789, the German chemist Klaproth discovered a new element. He named it 'uranium', in honour of the discovery of the planet Uranus by the astronomer Herschel. It is a melancholy reflection that in Greek mythology Uranus represents the Heavens—the giver of all life—whereas Man's first significant use of uranium was at Nagasaki and Hiroshima. However, poor Klaproth was not to know that when he chose the name.

20.91 Although uranium is continually emitting small amounts of invisible radiation, the appearance and general physical properties of the metal are not unusual in any way. It is a lustrous, silvery-grey metal, which will take a high polish, but which tarnishes fairly quickly in air. Uranium is a 'heavy' metal, with a relative density of $18 \cdot 7$. The pure metal, is both malleable and ductile, and commercially it can be cast and shaped by either rolling or extrusion. Chemically, uranium is very like tungsten, in that it forms stable, hard carbides. In fact, before the Second World War, attempts were made to use uranium in high-speed steel, as a substitute for tungsten. Apart from the demands for uranium as a nuclear fuel, these steels would not have survived, since they were less satisfactory than the orthodox alloys.

20.92 Since 1944, uranium has been notorious in the field of nuclear energy, first in the provision of 'fissionable material' in the atom bomb, and subsequently as a source of fuel for the 'atomic pile'. Atoms of all elements are built up from varying numbers of fundamental particles called electrons, protons, and neutrons. The uranium atom is a particularly large one, and, if a neutron is fired into certain uranium atoms, these atoms become unstable and disintegrate, with the release of a large amount of heat energy.

As a result of this fragmentation, simpler atoms are produced, as, for example, those of the inert gas krypton. More neutrons are also released, and these may enter nucleii of other uranium atoms, so that a 'chain reaction' is set up, leading to an atomic explosion. The rate of the chain reaction can be controlled by using rods of hafnium (20.52) to 'blot up' excess neutrons. Very roughly, this is the principle of the production of nuclear energy in an atomic pile.

20.93 *Plutonium.* This element is a sort of metallurgical Frankenstein's monster. It does not occur as a natural element, but is formed by the transformation of uranium in the atomic pile. It is a white metal which is fairly stable in dry air, but in moist air it 'rusts' rapidly.

Plutonium is used in nuclear weapons, and as a nuclear fuel. Since it emits powerful radiation, it is a very toxic substance, and can be handled only in 'glove boxes'.

Chapter Twenty-one
The Machinability of Metals

21.10 The ease with which a material can be machined depends largely upon the mechanical technique employed, and involves such aspects as the design of tools, methods of lubrication, and so on. It is not our purpose to attempt a discourse on this vast subject, about which complete textbooks are written, but rather to describe the effect which the microstructure of a metal has upon its general machinability.

21.11 Machining is a cold-working process, in which the cutting edge of the tool forms chips or shavings of the material being machined. During cold-working processes, heat is generated, because mechanical energy is being absorbed. In a machining process, heat arises from this source, as well as from energy dissipated by ordinary friction. In fig. 21.1, heat is being generated at X and Y by friction between the tool and the work-piece, and at A and B by distortion of the metal being machined. If the

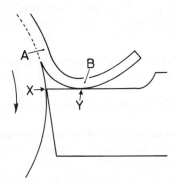

Fig. 21.1.

temperature at Y increases, this will prevent heat from being conducted away from the tip at X. Consequently, the temperature of the tip may rise to the extent that it softens, and loses its edge. For this reason, adequate lubrication and cooling, and the use of chip-breakers, are essential. Even so, for very high-speed cutting, the use of special tool materials (13.32 and 13.40) becomes necessary; whilst further improvements can sometimes be effected by using a work-piece of more suitable microstructure.

21.12 Very ductile alloys do not machine well, because local fracture does not occur easily under the pressure of the cutting-tool. Instead, such alloys spread under the pressure of the tool, and 'flow' around it, so that it becomes almost buried in the metal (fig. 21.2 (i)). Thus, considerable heat will be generated by friction. Not only is this a waste of energy, but loss of the tool edge is possible, as mentioned above.

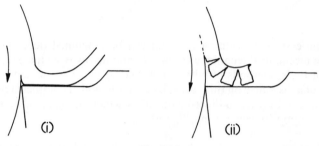

Fig. 21.2.

21.13 It follows that a brittle material will be far easier to machine than will a ductile one, since small chip-cracks form in advance of the cutting edge (fig. 21.2 (ii)), and there is less contact, and hence less friction, between tool and work-piece. Unfortunately, most brittle alloys are unsuitable for service in engineering, because they have little resistance to mechanical shock. However, it is possible to introduce *local* brittleness in an alloy, whilst at the same time producing only a relatively small reduction in the toughness of the material *as a whole*. This can be achieved in three ways:

(1) by the presence of a separate constituent in the microstructure,
(2) by suitable heat-treatment of the material before the machining operation, and
(3) by cold-working the material before the machining operation.

The presence of a separate constituent in the microstructure
21.20 Small isolated particles in the microstructure have the effect of setting up local stress concentrations as the cutting-tool travels towards them. It makes little difference whether the particles are hard or soft, since they behave as cavities in the structure, and act as stress raisers. Minute fractures therefore travel from the cutting edge to these particles, thus reducing friction between the tool and the material being cut (fig. 21.3). As well as reduction of wear on the tool, there will also be a reduction in the overall power required.

Moreover, due to the discontinuity introduced by the particles, the swarf will tend to fall away in small pieces, instead of forming the long, curly slivers associated with ductile materials.

21.21 Some alloys already have a constituent in the microstructure which puts them into the free-cutting category. Thus, slag particles in wrought iron, graphite flakes in grey cast iron, the hard intermetallic compound in high-tin bearing bronzes (16. 61), and the tiny particles of copper-aluminium

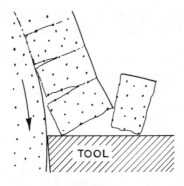

Fig. 21.3.

compound in annealed duralumin (17.72) put all these alloys into this free-machining class. In some materials, a deliberate addition is made, to provide isolated particles which will improve the machinability of a structure. Thus, in the cheaper free-cutting steels, the presence of a high sulphur content is utilised by adding a large excess of manganese, so that free globules of manganese sulphide are scattered throughout the structure. So, whilst a good-quality steel contains a maximum of 0·06% sulphur, a free-cutting steel may contain as much as 0·2% sulphur, along with 1·0% manganese, to ensure that all of the sulphur is present as isolated globules of manganese sulphide, and none as the brittle intercrystalline network of iron sulphide, which would form in the absence of manganese, and make the steel so brittle as to be useless.

21.22 Most copper used for electrical purposes is of high purity, since the presence of alloying elements or impurities generally leads to a big reduction in electrical conductivity. Unfortunately, the machinability of pure copper is very poor, mainly because it is such a ductile material. However, the addition of 1% tellurium to copper improves machinability considerability, yet reduces the conductivity by only 2%. The tellurium exists as finely dispersed particles throughout the structure of the copper.

21.23 The machinability of both ferrous and non-ferrous alloys is improved by the addition of small amounts of lead. Lead is insoluble in steels and most copper alloys, both in the solid *and liquid* states. It therefore exists as separate tiny spheres in the molten alloy, provided the latter is vigorously stirred to disperse the lead particles. When the alloy solidifies, the lead remains as isolated globules in the microstructure.

'Ledloy' steels contain no more than 0·2% lead, yet improvements in machinability of between 25 and 35% are reported. At the same time, there is very little deterioration in the mechanical properties due to the presence of the lead. The use of lead is particularly advantageous in introducing free-cutting properties into such alloys as the 13% chromium stainless steels, as the alternative method—utilising the presence of sulphur and manganese—may cut the impact toughness by half.

21.24 Lead has long been used to improve the machinability of brasses and bronzes, where relatively larger amounts in the region of 1·0% to 3·0% are used. An additional useful feature in the case of lead is that it acts as a lubricant during machining; the same is true of graphite, when machining a grey cast iron.

Class of steel	BS specification	Composition (%)		
Free-cutting mild steel	970: Part 1: 1972—220 M07	Carbon—0·1 *Manganese*—1·0 *Sulphur*—0·25		
Free-cutting mild steel	970: Part 1: 1972—240 M07	Carbon—0·1 *Manganese*—1·2 *Sulphur*—0·45		
Free-cutting '40' carbon steel	970: Part 1: 1972—212 M44	Carbon—0·40 *Manganese*—1·1 *Sulphur*—0·16		
Chromium rust-resisting steel (free-machining)	970: Part 4: 1970—416 S21 and 416 S41			Carbon—0·12 Chromium—13·0 *Manganese*—1·5 max. *Sulphur*—0·30 max.
		Optional but but total 1·0 maximum		{ *Selenium*—0·30 max. *Zirconium*—0·6 max. *Molybdenum*—0·6 max. *Lead*—0·35 max.
Free-cutting 18-8 stainless steel	970: Part 4: 1970—303 S21, 303 S41 and 325 S21	Carbon—0·1 Chromium—18·0 Nickel—10·0 *Manganese*—1·5 *Sulphur*—0·25 *Selenium*—0·25 Titanium—0·6		

Table 21.1—Typical free-cutting steels.

Suitable heat-treatment of the material before the machining operation
21.30 Coarse-grained materials are usually brittle, and so machinability often increases with grain size. However, this does not apply to very soft ductile materials, which are more easily machined when the grain size is small.

21.31 Low-carbon steels machine most easily in the normalised condition. Normalising produces small patches of pearlite, which break up the continuity of the soft ferrite, so that, during machining, they behave in much the same way, though to a lesser extent, as do the manganese sulphide globules in free-cutting steels.

21.32 Fine-grained steels have poorer machining characteristics than do coarse-grained steels of similar compositions. Machinability of these fine-grained steels can be improved by annealing them at a high temperature, in order to produce coarse-grained austenite, which, on cooling, will give rise to coarse-grained ferrite and pearlite. Other mechanical properties will suffer as a result of this treatment; so, after machining, the work will need to be normalised.

21.33 High-carbon steels are difficult to machine in the normalised condition. Machinability can be improved by giving a spheroidising anneal (11.52.2). This involves annealing the steel at a temperature just *below* the lower critical, i.e. at about 650°C, for about twenty-four hours. As a result of such treatment, the cementite layers in the pearlite 'ball up', and form separate globules dispersed throughout the ferrite. These globules assist machining in the usual way, but annealing should not be excessive, or very large cementite globules may form. Not only will these result in a rough machined surface, due to local tearing, but they may not dissolve again during subsequent heat-treatment process.

Cold-working the material before the machining operation

21.40 Materials supplied in rod or bar form for machining operations are generally cold-drawn or 'bright-drawn'. Free-cutting steels containing manganese sulphide are usually supplied as 'bright-drawn bar'. This treatment improves machinability by reducing the ductility of the surface layers, and introducing local brittleness there.

Chapter Twenty-two
Corrosion

22.10 Probably the reader's first serious confrontation with the problem of corrosion was when he became the proud possessor of his first second-hand motor car. What left the dealer's show-room as a 'mint-condition' vehicle began to show those ubiquitous rust spots within a few weeks. It is a melancholy reflection that, whilst the engine and any other worn mechanical parts of a motor car can be replaced almost indefinitely, once the body —which is also the 'chassis'—of a modern car begins to deteriorate badly, little can be done save consign the car to the scrap yard.

22.11 However, the corrosion of steel is a problem which in some degree must be faced by all of us, and it is estimated that the annual cost of the fight against corrosion on a world-wide basis runs into some thousands of millions of pounds. Other engineering metals corrode when exposed to the atmosphere, though generally to a lesser extent than do iron and steel. Because of their high resistance to all forms of corrosion, plastics are replacing metals for many applications where corrosion-resistance is more important than mechanical strength.

Chemical corrosion
22.20 Metals may corrode by a process which we can describe as simple chemical attack. Thus oxygen, ever present in the atmosphere, can combine with some metals to form a film of oxide on the surface. If this film is porous, or if it rubs off easily, the process of oxidation can continue, and the metal will gradually corrode away. Aluminium oxidises very easily, but fortunately the thin oxide film so formed is very dense, and sticks tightly to the surface, thus effectively protecting the metal beneath.

22.21 The rusting of iron and steel is not a case of simple oxidation, and is associated with the presence of both air *and moisture.* Iron will not rust in dry air, nor will it rust in pure water; but, when air and moisture are present together, iron, and particularly steel, will begin to rust very quickly. Rusting continues unabated, because the layer of corrosion product formed is loose and porous, so that a fresh film of rust will form beneath and lift upwards the upper layer.

Electrolytic corrosion
22.30 Essentially, this is also a form of chemical corrosion, though a little more complex in its origin than the simple chemical corrosion described above.

22.31 Most readers will be familiar with the principle of a simple electric cell. It consists of a plate of copper and a plate of zinc, both of which are immersed in dilute sulphuric acid. If the plates are not touching each other in the solution, and are not connected to each other outside the solution, then no action takes place; but, as soon as they are connected

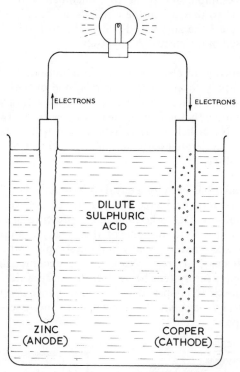

ELECTRONS   ELECTRONS

DILUTE
SULPHURIC
ACID

ZINC
(ANODE)

COPPER
(CATHODE)

Fig. 22.1—The 'corrosion' of zinc in the simple voltaic cell.

(fig. 22.1), a current of electricity, sufficient to light a small bulb, flows through the completed circuit. At the same time, bubbles of hydrogen form at the copper plate, whilst the zinc plate begins to dissolve in the acid, to form a salt, zinc sulphate. In this way, the 'chemical potential energy' of the zinc is being converted to electrical energy.

22.32 If the reader examines fig. 22.1, he will notice that the terminology used is opposite to that which may have been used in his school science lessons. The usual electrical terminology shows the zinc plate as being the negative pole (or cathode), whilst the copper plate is shown as being the positive pole (or anode). Similarly, the current is usually shown as flowing from copper to zinc in the external circuit, and not from zinc to copper as indicated in fig. 22.1. This apparent confusion arises from the fact that, in the early days of electricity, the true nature of an electric current was not really understood. It was thought that 'positive' electricity flowed from copper to zinc in the external circuit; whereas we now know that an electric current consists of a stream of *negatively-charged* particles (electrons) flowing from zinc to copper.

However, the important point to remember is that zinc is anodic towards copper; so that, when these metals are connected and immersed in any electrolyte,* the zinc will dissolve—or corrode—far more quickly than if immersed in the electrolyte by itself.

This phenomenon of electrolytic action is not confined to copper and zinc, but will apply to any pair of metals, one of which will always be anodic to the other. Often the rate of electrolytic action will be extremely slow, and consequently the flow of current between the two so small as to be unnoticed. Nevertheless, electrolytic action will be taking place, and will lead to the accelerated corrosion of that member of the pair which is anodic to the other.

22.33 If the reader has been long engaged in practical engineering, he may have learned that it is considered bad practice to use two dissimilar metals in close proximity to each other and in the presence of even such a weak electrolyte as rain-water. In spite of this, many examples of such bad practice are encountered, especially in domestic plumbing. Figure 22.2 (i) illustrates such a case. Here a section of mild steel pipe has been attached to one of copper, the system being used to carry water. Since the mild steel will be anodic to copper, it will rust far *more quickly than if the pipe were of mild steel throughout*. Rusting will be the more severe where it joins the copper pipe.

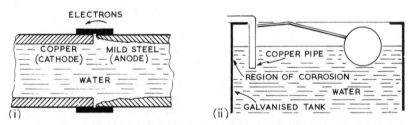

Fig. 22.2—Examples of bad plumbing practice.

22.34 A further example of bad practice, all too frequently encountered in a domestic installation, is shown in fig. 22.2 (ii). If the reader examines his cold-water tank, the chances are that he will find that the inlet pipe is of copper, whilst the tank is of galvanised (zinc-coated) mild steel. Since most mains water contains sufficient impurities to make it a mild electrolyte, the inevitable electrolytic corrosion will have taken place. As a

* For our purposes this means any solution which will conduct electricity.

result, the zinc coating will have corroded in an area adjacent to the copper inlet pipe; and the exposed mild steel, also anodic to copper, will also be rusting rapidly, as a result of the close proximity of the copper pipe. Enlightened plumbers will use a plastic inlet pipe.

22.35 Many pure metals have a good resistance to atmospheric corrosion. Unfortunately, these metals are usually expensive, and many of them are mechanically weak. However, a thin coating of one of these metals can often be used to protect mild steel. Pure tin has an excellent resistance to corrosion, not only by the atmosphere and by water, but by very many other liquids and solutions. Hence tinplate—that is, mild-steel sheet with a thin coating of tin—is widely used in the canning industry. Figure 22.3 illustrates what happens if a tin coating on mild steel becomes scratched.

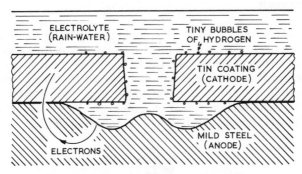

Fig. 22.3—The corrosion of scratched tinplate.

The mild steel is anodic to tin, and so it will corrode far more quickly in the region of the scratch than if the tin were not there at all. Consequently, to be of any use in protecting mild steel, a tin coating must be absolutely continuous and unbroken.

22.36 Figure 22.4 illustrates the effect of a zinc coating in similar circumstances. Here the zinc is anodic to the mild steel, and so in this case it is

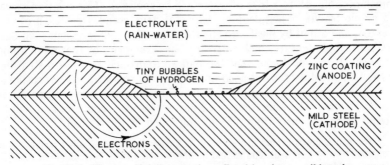

Fig. 22.4—The 'sacrificial' protection offered by zinc to mild steel.

the zinc which corrodes in preference to the mild steel. In fact, *the mild steel will not rust as long as any zinc remains in the vicinity of the original scratch.* Thus, by being dissolved itself, the zinc is in fact protecting the mild steel from corrosion. Such protection offered to the mild steel by the zinc is known as 'sacrificial protection'. Naturally the zinc will corrode fairly quickly under these circumstances, and the time for which this sacrificial protection lasts will be limited. For this reason, every attempt must be made to produce a sound coating, whatever metal is used; but, if the coating is anodic to the mild steel beneath, it will still offer some protection, even after the coating is broken. Thus galvanising is more useful than tin-coating in protecting steel. Tin-coating is used only in canning and in food-preparing machines generally, because it is not attacked by animal or fruit juices. Such juices would attack zinc coatings, and possibly produce toxic compounds.

22.37 The reader may verify the foregoing principles by means of a very simple experiment. Obtain three identical clean iron nails, a small strip of tin foil, and a strip of zinc sheet. The latter may be obtained from an old dry-battery, provided it is cleaned, and washed free of any electrolyte. To obtain tin foil is not so easy these days, since most metal foil used in food and cigarette packing is now of aluminium. Pierce two holes in each metal strip, and thread it on to a nail, as shown in fig. 22.5. Take three

Fig. 22.5.

separate glass containers, and fill each with a very weak salt solution (a pinch of table salt to a litre of water will be enough). This solution serves as an electrolyte. Then drop a nail/metal strip 'couple' into each of two vessels, and a nail by itself into the third vessel. Examine them each day, and note the extent to which corrosion has taken place by the end of about a week. It will be found that the nail accompanied by the tin strip has rusted most; whilst the one accompanied by the zinc strip has not rusted at all. Some white corrosion product from the zinc will, however, be found to be present. The unaccompanied nail serves as a 'control', and will be found to be less rusty than that accompanied by the tin foil.*

* The reader could extend this experiment further by including other metal foils in conjunction with iron nails, and so find to what extent a metal was anodic or cathodic towards mild steel.

There is always a danger of electrolytic corrosion taking place in the steel hull of a ship, due to the proximity of the bronze propellor (steel is anodic to bronze, and sea water is a strong electrolyte). However, if a zinc slab is fixed to the hull, near to the propellor, this will sacrificially protect the steel hull, since zinc is more strongly anodic than steel in this instance. Naturally, the zinc slab must be replaced from time to time, as it is used up.

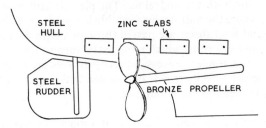

Fig. 22.6—Sacrificial protection of a ship's hull by zinc slabs.

The protection of metal surfaces
22.40 The corrosion of carbon steel is due partly at least to electrolytic action between different phases in the structure of the steel. Pearlite consists of microscopically thin layers of ferrite and cementite, arranged alternately in the structure (11.43). Ferrite is anodic to cementite, and so it is this ferrite which corrodes away, leaving the cementite layers standing proud. Being very brittle, these cementite layers soon break away.

In order to protect the surface of steel, therefore, it must be coated with some impervious layer which will form a mechanical barrier against any electrolyte which is likely to come into contact with its surface. In all cases, the surface to be coated must be absolutely clean and rust free.

22.41 *Painting* is used to coat vast amounts of mild steel, not only to protect it against corrosion by the atmosphere, but to provide an attractive finish. Optimum results are obtained by first 'phosphating' the surface of the steel. This involves treating it with a phosphoric acid preparation, which not only dissolves rust, but also coats the surface of the steel with a dense and slightly rough surface of iron phosphate. This affords some protection against corrosion, but also acts as an excellent 'key' for the priming paint and the undercoat of subsequent paint.

22.42 *Stove-enamelling.* This finish is used to provide a hard-wearing corrosion-proof coating for many domestic appliances, such as washing-machines, refrigerators, cooking-stoves, and the kitchen sink.

22.43 *Coating the surface with another metal.* A thin coating of a corrosion-resistant metal can be applied to one which is less corrosion-resistant, in order to protect it. The aim is *always* to provide a *mechanical* barrier against possible electrolytes or corrosive atmospheres, but it must be remembered that, whilst zinc and aluminium will offer sacrificial protection should coatings of these metals become damaged, the presence of damaged coatings of most other metals will *accelerate* corrosion.

The metallic coating can be applied in a number of different ways.

22.43.1 *Hot-dipping* can be used to coat the surface of iron and steel components with both tin and zinc. Tin plate is still manufactured in South Wales, where the industry was established some three hundred years ago. Clean mild-steel sheets are passed through a bath of molten tin, and then through squeeze rolls, which remove the surplus tin.

Galvanising is a similar process, whereby articles are coated with zinc. Buckets, dustbins, wheelbarrows, cold-water tanks, and barbed-wire are all coated by immersion in molten zinc.

22.43.2 *Spraying* can be employed to coat surfaces with a wide range of molten metals, though zinc is most often used. In the Schoop process, an arc is struck between two zinc wires within the spray gun, and the molten metal so produced is carried forward in an air blast. This type of process is useful for coating structures *in situ*, and was employed in the protection of the Forth road-bridge.

22.43.3 *Sherardising* is a 'cementation' process, similar in principle to carburising. Steel components are heated in a rotating drum containing some zinc powder at about 370°C. A very thin but uniform layer of zinc is deposited on the surface of the components. It is an ideal method for treating nuts and bolts, the threads of which would become clogged during ordinary hot-dip galvanising.

22.43.4 *Electroplating* is used to deposit a large number of metals on to both metallic and non-metallic surfaces. Gold, silver, nickel, chromium, copper, cadmium, tin, zinc, and some alloys can be deposited in this way. Electroplating is a relatively expensive process, but provides a very uniform surface layer of very high quality, since accurate control of the process is possible at all stages. Moreover, heating of the component being coated is not involved, so that there is no risk of destruction of mechanical properties which may have been developed by previous heat-treatment.

22.43.5 *Cladding* is applicable mainly to the manufacture of 'clad' sheet. The basis metal is sandwiched between sheets of the coating metal, and the sandwich is then rolled to the required thickness. During the process, the coating film welds on to the base metal. 'Alclad', which is duralumin coated with pure aluminium, is the best known of these products.

22.44 *Protection by oxide coatings.* In some cases, the film of oxide which forms on the surface of a metal is very dense, and closely adherent. It then protects the metal beneath from further oxidation. The 'blueing' of ordinary carbon steel during tempering produces an oxide film which offers some protection against corrosion.

22.44.1 *Anodising* is applied to suitable alloys of aluminium, in order to give them added protection against corrosion. The natural oxide film on the surface of these materials is an excellent barrier to further oxidation, and, in the anodising process, this film is thickened by making the article the anode in an electrolytic bath. As current is passed through the bath, atoms of oxygen are liberated at the surface of the article, and these immediately combine with the aluminium, thus thickening the natural oxide film. Since aluminium oxide (alumina) is extremely hard, this film is also wear-resistant; it is also thick enough to enable it to be dyed an attractive colour.

Metals and alloys which are inherently corrosion-resistant
22.50 Stainless steels are resistant to corrosion partly because the tenacious chromium oxide film which coats the surface behaves in much the same way as does the oxide film on the surface of aluminium. They are also corrosion-resistant, because of the uniform structure (13.51) which is generally present in such steels. If a structure consists of crystals which are *all of the same composition*, then there can be no electrolytic action between them, as there is, for example, between the ferrite and cementite in the structure of an ordinary carbon steel.

Pure metals are generally corrosion-resistant for the same reason; though, of course, particles of impurity present at crystal boundaries can give rise to intercrystalline corrosion. For this reason, extremely pure metals have very high resistances to corrosion. This applies particularly to iron and aluminium, which can be obtained as much as 99·9999% pure. Unfortunately, such metals are generally so expensive in this state of purity that their everyday use is not possible.

Chapter Twenty-three
Plastics

23.10 The commercial production of polythene—a British invention—began on the day in early September 1939 when Hitler's forces marched into Poland, precipitating the Second World War. One of the early uses of polythene was as an insulation material in the cables used in radar, a factor so vital to us in attaining a successful conclusion to the war. Nylon, now one of the important 'man-made fibres', was developed in the USA in the years just before the war, but made its glamorous début to the public in general in the form of ladies' stockings in the spring of 1940. Both of these materials are typical of those generally referred to as 'plastics', but it must be appreciated that the plastics industry was in fact founded in much earlier times.

23.11 In the wider sense, any substance which can be shaped by moulding in the solid state by the application of pressure can be termed a plastic. Thus, resins, clay, and sealing-wax were used quite early in Man's history. The technology of rubber began in 1820, when a British inventor, Thomas Hancock, developed a method for shaping raw rubber; whilst some twenty years later, in the USA, Charles Goodyear established the 'vulcanisation' process by which raw rubber was made to 'set' and produce a tough, durable material, later to be so important in the growth of the motor-car industry.

Shortly afterwards, the plastics cellulose nitrate and celluloid were developed from ordinary cellulose fibre, and when, in the early years of this century, Dr Leo Baekeland, a Belgian chemist, introduced the substance 'bakelite', the plastics industry could be said to have 'arrived'.

23.12 At present, the use of plastics continues to increase rapidly, as new materials are developed. A popular make of British car contains some 20 kg of plastic components. These include the radiator grille, a material being chosen which is proof against both corrosion and flying stones. A spokesman of the company said recently, 'There is no doubt that well before the end of the century, mass-produced plastic [car] bodies will be commonplace'. Those of us who have presided over the rusting away of a succession of expensive motor cars will be grateful for this promise.

Types of plastics
23.20 Bakelite differs from either polythene or nylon in one important respect. Whereas the latter substances will soften repeatedly whenever they are heated to a high enough temperature, bakelite does not. Once moulded

to shape, it remains hard and rigid, and reheating has no effect on it—unless the temperature used is so high as to cause it to decompose.

23.21 Plastics can therefore be classified into three groups.

(1) Thermoplastic materials—substances which lose their rigidity whenever they are heated, so they can be moulded repeatedly.

(2) Thermosetting materials, which undergo a definite chemical change during the moulding process, causing them to become permanently rigid, and incapable of being softened again.

(3) Cold-setting plastics, which become permanently hard, due to a chemical reaction which occurs at ordinary temperatures. Materials used in conjunction with glass-fibre for the repair of the bodywork of decrepit motor cars fall into this class.

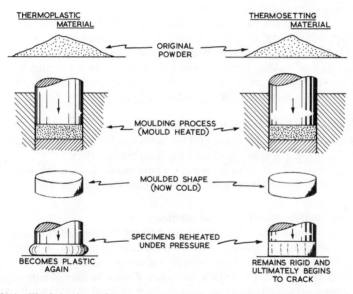

Fig. 23.1—The behaviour of thermoplastic and thermosetting materials when reheated under pressure.

23.22 The raw materials used in the manufacture of plastics can be divided into three main classes.

(*a*) Animal or vegetable products, which include casein obtained from cow's milk, and cellulose obtained principally from cotton fibre too short for spinning, and from wood pulp.

(*b*) Coal by-products, obtained during the distillation of coal to produce gas.

(*c*) Petroleum by-products, obtained during the refining and 'cracking' of crude oil.

General properties of plastics

23.30 When visiting the wild and more remote parts of Europe's coast-line, one finds, cast up by the tide, along with the seaweed and driftwood, a motley collection of plastic bottles which once held detergent liquids. Because of its low relative density and comparative indestructibility, this plastic junk will presumably congregate in increasing quantity, as yet another example of Man's careless pollution of his environment.

23.31 This fact at least illustrates some of the more important properties of plastics generally.

(1) They are resistant both to atmospheric corrosion and to corrosion by many chemical reagents.

(2) They have a fairly low relative density—a few will just float in water, but the majority are somewhat more dense.

(3) Many are reasonably tough and strong, but the strength is less than that of metals. However, since the relative density of plastics is low, this means that many have an excellent strength/weight ratio. One feels that in this direction the properties of plastics have yet to be fully exploited.

(4) Most of the thermoplastic materials begin to soften at quite low temperatures, and few are useful for service at temperatures much above 100°C. Strength falls rapidly as the temperature rises.

(5) Most plastics have a pleasing appearance, and can be coloured if necessary. Some are transparent and completely colourless.

The composition and structure of plastics

23.40 All materials included under our present meaning of the term 'plastics' are 'organic' compounds, based on the element carbon. Organic compounds are associated mainly with living matter, and those which occur naturally are of animal or vegetable origin. Cellulose fibre can be regarded as the material which constitutes the skeleton of most plants; whilst petroleum and coal, from the by-products of which most of our plastics are derived, were produced by the decay and fossilisation of vegetable matter, millions of years ago. In addition to the element carbon, most plastics contain hydrogen, whilst many contain oxygen. A smaller number contain other elements, the chief of which are nitrogen, chlorine, and fluorine.

23.41 All organic substances exist in the form of molecules. In these molecules, atoms are bound together by strong forces of attraction. Possibly the simplest organic molecule is that of the gas methane (North Sea gas). Such a molecule consists of four atoms of hydrogen bound to an atom of carbon, something like this:

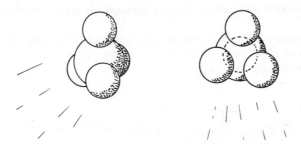

23.42 In addition to the strong binding forces between atoms in a molecule, much weaker forces of attraction operate between the molecules themselves. When the molecules are simple and small, as in the case of methane, these forces of attraction are small. Hence methane remains a gas at normal temperatures and pressures.

23.43 Plastics, however, consist of a mass of very large molecules, each molecule containing several thousand atoms tightly bound to each other. Moreover, since these molecules are very large, the forces of attraction between them will be considerable. The actual shape of the molecule also affects the properties of a plastic. In thermoplastic materials, the carbon atoms are attached to each other in the form of a long chain—thus, in polythene, a chain of about 1200 carbon atoms in length is formed (shown black in the diagram), the hydrogen atoms being attached to individual carbon atoms:

Since these molecules are so large, much greater forces of attraction operate between them, particularly when they lie parallel and very close to each other. Furthermore, it is easy to imagine that a considerable amount of entanglement will exist among these long chain-like molecules:

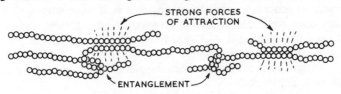

23.44 Strong binding forces operating between the molecules, as suggested above, will give rise to the formation of a solid possessing both strength and rigidity. When such a solid is heated, the distances between aligned molecules become greater, and so the forces of attraction between them will decrease.* Then the material will be weaker and less rigid, so that it

* In a similar way the force acting between two magnets becomes smaller the further they are moved apart.

can be moulded more easily. Such a substance is said to be *thermoplastic*.

23.45 In some plastics, a chemical change is initiated during the moulding process, and 'cross-links', in the form of strong chemical bonds, are formed between adjoining molecules:

ATOMS FORMING STRONG CHEMICAL BONDS

The powerful forces associated with these chemical bonds are much greater than the simple forces of attraction operating between the chain molecules in a thermoplastic substance. Consequently, the molecules are unable to slide over each other, and a strong rigid three-dimensional network is formed. Thus, *thermosetting* has taken place.

23.46 In natural rubber, the long-chain molecules are more complex in form than are those in polythene, and consist of chains 44 000 or more carbon atoms in length. Moreover, these rubber molecules are kinked and folded, so they possess elasticity in a manner similar to that of a coiled spring. At the same time, natural rubber stretches like dough when stressed, because the chain molecules slide past each other, into new positions (fig. 23.2). For this reason, a piece of natural rubber is both elastic and plastic at the same time, and, when stretched, it will not return to its original length.

When up to 3% sulphur is added to rubber before it is moulded, 'vulcanisation' will occur, if the moulding temperature is high enough. Atoms of sulphur form powerful links between the rubber molecules, so that they are permanently anchored to each other. Thus, whilst the vulcanised rubber retains its elasticity, due to the folded nature of its molecules, it retains its shape permanently, because of this cross-linking. If the vulcanisation process is carried further—that is, if more sulphur is used—so many cross-links are produced between adjoining molecules that the whole mass becomes rigid, and loses most of its elasticity. The product is a black plastic known as 'ebonite' or 'vulcanite', and was used for the manufacture of such articles as fountain-pens, before other more suitable plastics were developed.

Some terms used in plastics technology
23.50 The household plastic we call 'polythene' is more correctly known as 'polyethylene'. It is made from the gas ethylene, a by-product of the petroleum industry. In the ethylene molecule, a weak 'double-bond' exists between two carbon atoms (fig. 23.3 (i)), and suitable chemical treatment causes this bond to break. When this occurs simultaneously amongst

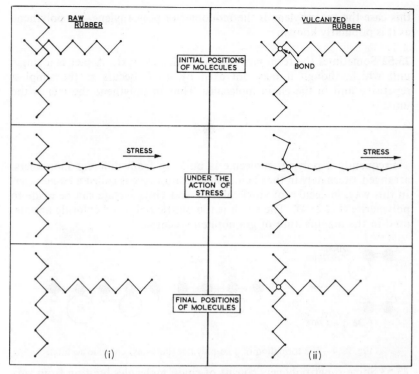

Fig. 23.2—Rubber molecules are of the long-chain type.

Due to their 'folded' form, they become extended in tension, but return to their original shapes when the stress is removed. In raw rubber (i), a steady tensile force will cause separate molecules to slip slowly past each other, into new positions; so, when the force is removed, some plastic deformation remains, although the elastic deformation has disappeared. By 'vulcanising' the raw rubber (ii), the chain molecules are 'tacked' together at certain points, so that no permanent plastic deformation can occur, and only elastic deformation is possible.

many molecules of the gas, the resultant units (fig. 23.3 (ii)) are able to link up, forming long-chain molecules of polyethylene (fig. 23.3 (iii)).

Fig. 23.3—The polymerisation of ethylene to polyethylene ('Polythene').

23.51 This type of chemical process is known as *polymerisation*, and the product is called a *polymer*. Since many of the chain molecules so produced are extremely long, the term *super-polymer* is often applied. The simple substance from which the polymer is derived is called a *monomer*; so in this case the gas ethylene is the monomer of polyethylene—or polythene as it is popularly known.

23.52 Sometimes the term *mer* is used in this context. A mer is a single unit which, though it may not exist by itself, occurs as the simplest repetitive unit in the chain molecule. Thus in polythene the mer is the unit:

Frequently a polymer molecule is built up from different monomers, arranged alternately in the chain. Such a substance is called a *co-polymer*. In this way, molecules of vinyl chloride and vinyl acetate can be made to polymerise (fig. 23.4). The result is the plastic polyvinyl chloride acetate, used in the manufacture of gramophone-records.

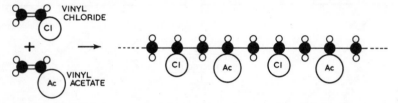

Fig. 23.4—The formation of a co-polymer (polyvinyl chloride acetate).

23.53 Some super-polymers consist of chain molecules built up from very complex monomers. This often leads to the operation of considerable forces of attraction between the resulting molecules at points where they lie alongside each other; consequently, such a material lacks plasticity, even when its temperature is increased. Cellulose is a natural polymer of this type.

23.54 In 1846, Dr Frederick Schönbein, of the University of Basle, produced cellulose nitrate by treating ordinary cellulose with nitric acid. When heated, it proved to be slightly more plastic than ordinary cellulose. This is because the 'nitrate' side branches (fig. 23.5 (ii)) which had been attached to the chain molecules as a result of treatment with nitric acid, act as 'spacers', separating the chain molecules, and so reducing the forces of attraction between them.

In 1854, Alexander Parkes, the son of a Birmingham industrialist, began to experiment with cellulose nitrate, with the object of producing a mouldable plastic from it. He found that, by adding some camphor to cellulose nitrate, a mixture was produced which passed through a mouldable stage when hot. Not long afterwards, this substance was produced commercially, under the name of 'celluloid'. The bulky camphor molecules separate the chain molecules of cellulose nitrate by even greater distances, so that the forces of attraction between them are reduced further. Thus the plasticity

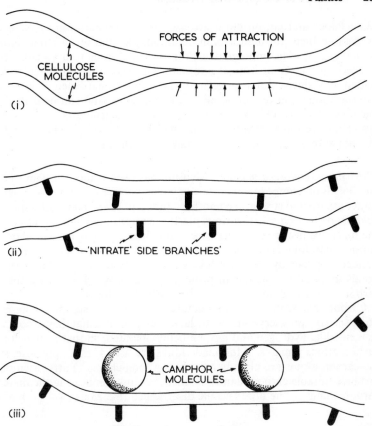

Fig. 23.5—The function of a plasticiser.

of celluloid is greater than that of cellulose nitrate. In this instance, camphor is termed a *plasticizer* (fig. 23.5 (iii)). Celluloid is a very inflammable material, and this fact has always restricted its use. Before other plastic materials were freely available, much celluloid was used in the manufacture of dolls and other children's toys, often with tragic results when these articles were brought too near a fire. An even earlier use for celluloid was as a replacement for ivory in the manufacture of billiard balls.

23.55 Many plastics, particularly those of the phenolic ('bakelite') type, are rather weak and brittle in the moulded state. Fortunately, these mechanical properties can be considerably improved by introducing some fibrous material to the powder being moulded. Such materials are called *fillers*, and include wood-flour, waste sawdust, paper, rag, and cotton fibre. Fillers are also used for specific purposes. Thus, asbestos fibre (25.50) improves resistance to high temperatures, whilst mica increases electrical resistivity.

23.56 Paper and cotton cloth are used in the preparation of *laminates*, by passing them through a bath containing a solution of some suitable phenolic resin. The impregnated cloth is then allowed to dry before being stacked, one layer on top of another. The resulting 'sandwich' is then heated and squeezed in a press until the resin has hardened. These laminates are tough, have a good electrical resistance, and can be drilled and machined successfully. Wood veneers can be incorporated in the surface layers, the resultant product being widely used for panelling, and the manufacture of furniture, radio cabinets, and the like.

23.57 *Fibre-reinforced plastics.* Mention has been made elsewhere (23.62 and 25.50) of composite materials in which particles or fibres of one substance are used to strengthen another. In recent years, many attempts have been made to strengthen various plastics with suitable fibres, and possibly the most promising results have been obtained by using carbon filaments, consisting of long chains of carbon atoms. These carbon atoms are firmly bonded together by electrical forces acting between them. The chemical bonds involved are similar in principle to those which bind together the carbon atoms in diamond, giving it its great hardness.

Carbon filaments are manufactured by heating suitable man-made fibres (usually polyacrylonitrile) to such a temperature that they decompose, elements like hydrogen and oxygen being liberated, so that a 'skeleton' of carbon atoms remains. Man-made fibres are used for this purpose, since the carbon atoms are already arranged in long chains. Early attempts to produce suitable carbon filaments failed, because, on heating the material, carbon chains in the fibre became disarranged (fig. 23.6 (i)).

(i)   (ii)

Fig. 23.6.

This problem was overcome by heating the original fibres whilst they were in tension. Under these conditions, the man-made fibre threads decompose, but the carbon chains in the filaments which remain are kept in the desired position (fig. 23.6 (ii)).

The product of this operation is a bundle of carbon filaments looking very much like black silk in fine fibre form. This material has a very high tensile strength, of the order of 2900 N/mm^2. However, it is of little use in this condition, since it possesses no rigidity. To produce a strong, rigid material, bundles of these fibres are placed in a suitable mould, and a cold-setting resin mixture is poured over them (fig. 23.7). When the resin sets, a very light-weight material of high strength and rigidity is produced.

Bundles of these carbon filaments are used to strengthen the hulls of small plastic racing-boats and kayaks, as well as the plastic bodywork of

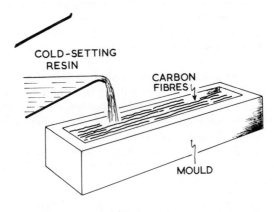

Fig. 23.7.

some racing-cars. A considerable saving in weight is obtained in both cases, due to the very high strength/weight ratio of the carbon fibre/plastic composites. The use of these materials will undoubtedly increase in the future.

Some important plastics
It is convenient to divide these materials into two main groups—those which are thermoplastic, and those which are thermosetting.

23.60 Thermoplastic materials
23.61 *Cellulose-type plastics* are derived from natural cellulose, one of the world's most plentiful raw materials. It occurs in many forms of plant life, but much of the raw cellulose used in the plastics industry is as cotton linters (cotton fibres too short for spinning to yarn).

23.61.1 *Cellulose nitrate* has already been mentioned (23.53) as the basis for the manufacture of celluloid. Whilst some is still used in the moulded form, the bulk of cellulose nitrate is now used for making laquers, 'leather' cloth, and thin sheets.

23.61.2 *Cellulose acetate* is the most important of the modern cellulose-base plastics. It was used many years ago in the manufacture of safety-glass, in which a layer of cellulose acetate was sandwiched between two sheets of glass. However, its one great advantage over cellulose nitrate is that, unlike the latter, it is not explosive, but is virtually non-inflammable. Consequently, it has largely replaced cellulose nitrate as a base for ciné-film and photographic film generally. Although cellulose acetate plastic is not particularly tough or strong, it is fairly cheap, and is used for mould-ing a wide variety of articles, from pens and pencils to radio cabinets.

23.61.3 *Cellulose acetate butyrate* is more resistant to moisture than is ordinary cellulose acetate, and at the same time retains the other useful properties of cellulose acetate. Because of its greater resistance to moisture, it is useful for the manufacture of handles for brushes and cutlery.

23.62 *Vinyl plastics* are based on vinyl compounds, some of which are of the self-linking variety; that is, they will undergo polymerisation of their own accord, monomers linking together spontaneously at ordinary temperatures, to form polymers. Some of the vinyl compounds begin as watery liquids, but, if allowed to stand for some time, they become more and more viscous as polymerisation proceeds, ultimately attaining a solid, glass-like state. They are thermoplastic, and become soft and mouldable again when warmed.

SHAPE OF MOLECULE   VISCOSITY OF MATERIAL

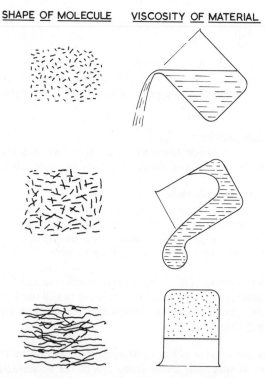

Fig. 23.8—Stages in the spontaneous polymerisation of some vinyl compounds.

23.62.1 *Polyvinyl acetate* softens at too low a temperature for it to be of much use as a mouldable plastic, but it is useful as an adhesive, since it will stick effectively to almost any surface. It is supplied as an emulsion with water; when used in the manner of glue, the water dries off, leaving a strong, adhesive polyvinyl acetate film for bonding the parts together.

23.62.2 *Polyvinyl chloride.* The gas vinyl chloride was discovered more than a century ago. It was found that, on heating, it changed to a hard, white solid, later identified as polyvinyl chloride. Having rather a high softening temperature, PVC was difficult to mould, and it was not until the late nineteen-twenties that it was discovered that it could be plasticised. By adjusting the proportion of liquid plasticiser used, a thermoplastic material can be produced which varies in properties from a hard rigid substance to a soft rubbery one. Consequently, during the Second World War, and since then, PVC has been used in many instances to replace rubber. Protective gloves and garden hose are examples of such use. PVC can be shaped by injection moulding, extrusion, and the normal thermoplastic processes. It can also be compression-moulded without a plasticiser, to give a tough, rigid material such as is necessary in miners' helmets.

23.62.3 *Polyvinyl chloride acetate* co-polymers have already been mentioned (23.52). In these, the very high softening temperature of PVC is reduced by combining vinyl chloride with vinyl acetate, which has rather a low softening-temperature when polymerised to polyvinyl acetate.

23.62.4 *Polyvinylidene chloride* is polymerised from vinylidene chloride, which is similar to vinyl chloride, except that its molecule contains an extra atom of chlorine. By making a co-polymer of vinylidene chloride and vinyl chloride, an extremely tough, corrosion-resistant material is produced. A coarse fabric, woven from thin extruded strips, is used for upholstery, particularly in the motor car, and in garden furniture which must withstand the weather.

23.62.5 *Polyethylene* or 'polythene' is possibly the best known of the thermoplastic materials. It has an excellent resistance to corrosion by most common chemicals, and is unaffected by foodstuffs. Polythene is tough and flexible, has a high electrical resistance, is light in weight, and is easily moulded and machined. Since it is also very cheap to produce, it is not surprising that polythene finds such a wide range of applications.

23.62.6 *Polypropylene* is generally similar in its properties to polythene, though it has better heat-resistance. Whilst polythene is polymerised from the gas ethylene, polypropylene is polymerised from the gas propylene. Development of polypropylene is still progressing, and currently large quantities are being used for bodywork in the automobile industries.

23.62.7 *Polystyrene* was developed in Germany before the Second World War, and is made by the polymerisation of styrene, a chemical substance known since 1830. Polystyrene has now become one of the most important of modern thermoplastic materials. It is a glassy, transparent material, similar in appearance to PMM (23.67), and is strong, flexible, and light in weight. Polystyrene hollow-ware is fairly easy to identify as such, since it emits a resonant 'metallic' note when tapped smartly. One most important feature it its high electrical resistance. Polystyrene can be cast, moulded, and extruded; and easily shaped by the ordinary hand methods of sawing, filing, and drilling. Mouldings in polystyrene are extremely tough at low temperatures, for which reason they are particularly useful as refrigerator parts.

A number of plastics can be made into foams by causing gas bubbles to be released within the plastic during the setting process. Foamed (or 'expanded') polystyrene is made with up to 97% of its total volume consisting of air. It then has a lower relative density than either balsa-wood or cork, and has an important advantage over the latter in that it does not absorb water when used as a flotation material. Expanded polystyrene is used for the manufacture of ceiling tiles, which, because of the large proportion of air bubbles they contain, provide excellent thermal insulation. It is also familiar as a packaging material for fragile equipment such as cameras and transistor radios.

23.62.8 *ABS plastics* are so called because they are co-polymers of acrylonitrile, butadiene, and styrene. They have an outstanding resistance to fracture by impact, and also a high tensile strength. They are resistant to acids and alkalis, as well as to some organic solvents. ABS plastics are available in the form of moulding powders for injection-moulding and extrusion; sheet is also available for vacuum-forming processes.

Large amounts of ABS are now used in the automobile industry, where glass-fibre reinforced plastics are finding increasing use for panel work. Since both ABS and polypropylene can be chromium plated, these materials are now widely used for radiator grilles and similar components. On one continental car, no less than twenty-six parts are of chromium-plated plastics. Several cars now use bodywork composed almost entirely of ABS plastic laminates, whilst at least one sports car uses plastic reinforced with the relatively new carbon fibre.

26.63 *Fluorocarbons.* The most important of these materials is polytetrafluoroethylene—or PTFE, as it is commonly called. It can be regarded as polythene in which all of the hydrogen atoms have been replaced by atoms of the extremely reactive gas fluorine.* Since fluorine is a very reactive element, many of its compounds, once formed, are difficult to decompose. Thus, PTFE will resist attack by all solvents and corrosive chemicals; for example, hot, concentrated sulphuric acid does not affect it, whilst

* The gas fluorine is one of the same group of chemical elements as chlorine, a reactive poisonous gas used on the Western Front in the First World War. Fluorine is the most reactive of all elements, and will even attack glass.

aqua regia (a mixture of concentrated hydrochloric and nitric acids, which will dissolve platinum and gold) leaves PTFE unharmed. It will also withstand higher temperatures than most plastics, and can be used at temperatures up to 300°C. It is an excellent electrical insulator, and has a very low coefficient of friction. The high cost of fluorine—and hence of PTFE—at present restricts the use of this plastic, except where its particular properties are utilised.

23.64 *Polyamides* include a number of compounds better known under the collective name of 'nylon'. The most widely used of these compounds is 'nylon 66'. It is a strong, hard-wearing material, best known for its use in ladies' hosiery, but it must not be forgotten that it is also an important engineering material. In addition to its strength and toughness, nylon has a relatively high softening-temperature; so it can safely be heated in boiling-water. Nylon moulding powders can be shaped by injection-moulding and by extrusion.

23.65 *Polyesters* are sometimes of the setting variety (23.72), but one important member of the group is thermoplastic. This is polyethylene terephthalate—better known as 'Terylene'. These materials are made in a similar manner to the polyamides, producing long thread-like molecules. Terylene was developed in Britain during the Second World War, and has since become well established as the basis of a branch of the synthetic fibre industry.

23.66 *Acetal resins* possess a range of mechanical properties which are intermediate between those of metals and the more common plastics. Since they have a high tensile strength, coupled with toughness and rigidity, it is possible to use these materials to replace metal die-castings for many purposes. Dimensional stability and resistance to many solvents are also good.

23.67 *Acrylics* are a group of vinyl plastics of which the most important is polymethyl methacrylate (PMM). It is a clear, glass-like plastic, better known as 'Perspex' (in Britain) or 'Plexiglas' (in the USA), and was developed during the Second World War, for use in aircraft. Not only is it much tougher than glass, but it can easily be moulded. It is produced by the polymerisation of methyl methacrylate.

Since it will transmit more than 90% of daylight, and is much lighter and tougher than glass, it can be used in the form of corrugated sheets, interchangeable with those of galvanised iron, for use in industrial buildings. Lenses can be made from PMM by moulding from powder. This is an inexpensive method of production, as compared with the grinding and polishing of a glass blank. Unfortunately, PMM lenses scratch very easily.

Group	Compound	Relative density	Tensile strength (N/mm^2)	Chemical resistance
Cellulosics	Cellulose nitrate	1·37	40	Fair
	Cellulose acetate	1·30	33	Fair
Vinyls	Polyvinyl chloride (PVC)	1·35	55	Good
	Polyvinyl chloride acetate	1·3	25	Fairly good
	Polyvinylidene chloride ('Saran')	1·68	28	Very good
	Polythene	0·92	13	Excellent
	Polypropylene	0·90	33	Very good
	Polystyrene	1·06	45	Fairly good
	ABS	1·01	35	Very good
Fluorocarbons	Polytetrafluoroethylene (PTFE, 'Teflon')	2·15	25	Excellent
Polyamides	Nylon 66	1·12	Moulded—60 Filaments—350	Good
Polyesters (thermoplastic)	'Terylene', 'Dacron'	1·38	175	Moderate
Acetal resins		1·42	70	Fairly good
Acrylics	Polymethyl methacrylate (PMM, 'Perspex', 'Plexiglas')	1·18	55	Fairly good

Table 23.1—Properties and uses of thermoplastic materials.

afe orking mperature C)	Relative cost	Typical uses
5	Low	Brush- and cutlery-handles, drawing-instruments, fountain-pens, instrument-dials, labels, spectacle-frames, piano-keys, toilet-seats, tool-handles. Inflammability limits use
0	Low	Artificial leather, brush-backs, combs, spectacle-frames, photographic film, mixing-bowls, lamp-shades, toys, laminated luggage, knobs, wire- and cable-covering
5	Low	Artificial leather cloth, gloves, belts, toys, dolls, curtains, raincoats, safety-helmets, protective clothing, packaging, industrial piping, beading, radio components, cable-covering, factory-ducting, plating-vats
0	Moderate	Gramophone-records, containers, chemical equipment, screens, protective clothing
0	Moderate	Food-wrapping, packaging, filaments used in weaving upholstery fabrics, rugs and carpets, outdoor furniture, seats for buses, filters, chemical equipment, screens, coating of wire
5	Moderate	Acid-resisting linings, babys' baths, bottle-caps, dust-bins, electrical insulation, kitchen equipment, packaging-film, wrapping material, chemical equipment, cold-water plumbing, squeeze-bottles, surgical items, etc.
0	Moderate	Packaging, pipes and fittings, cable-insulation, battery-boxes, refrigerator parts, sterilisable bottles and other uses where boiling water is involved, cars
0	Low	Radio and TV parts, food-containers, kitchen equipment, refrigerator parts, toys, toilet articles. The foam form is used for ceiling tiles and packaging
0	Moderate	Pipes, radio-cabinets, tool-handles, valves, protective helmets, textile bobbins, pumps, battery-cases, luggage, Now used in large amounts for automobile bodywork
0	Very high	Gaskets, valve-packings, inert laboratory equipment, chemical plant, piston-rings, bearings, non-stick coatings (frying-pans), electrical insulation, filters
0	Fairly high	Raincoats, yarn (clothing), containers, cable-covering, gears, bearings, cams, spectacle-frames, combs, bristles for brushes, climbing-ropes, fishing-lines, shock-absorbers
5	Low	Synthetic fibres—a wide range of clothing
5	Moderate	Bearings and cams, gears, flexible shafts, office machinery, carburettor parts, pump-impellers and bowls, car instrument-panels, knobs and handles
5	Moderate	Aircraft glazing, building-panels, roof-lighting, baths, sinks, protective shields, advertising-displays, windows and windscreens, automobile tail-lights, lenses, toilet articles, dentures, knobs, telephones, aquaria, double-glazing, garden cloches

23.70 Thermosetting materials

23.71 *Phenolic plastics* are possibly the best known of the thermosetting materials.

23.71.1 *Phenol formaldehyde*, the original 'bakelite', is a member of this group, and is formed when phenol (or carbolic acid) reacts with the gas formaldehyde. There are three different points on a phenol molecule where links with other molecules can be produced, so that the molecule can be represented thus:

If phenol is mixed with a *limited* amount of formaldehyde, then molecules of each can react to form a long-chain molecule thus:

where the link ● is derived from the formaldehyde molecule. In this condition, the material is brittle but thermoplastic, and can be ground to a powder suitable for moulding. This powder (or 'novolak') contains other materials, which, on heating, will release more formaldehyde, so that, during the final moulding process, cross-links are formed between the chain molecules:

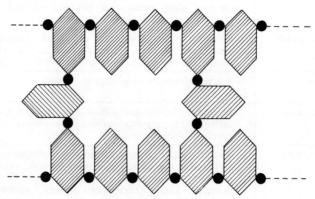

In this way, a rigid, three-dimensional network is formed, and the material has 'set' permanently. Since this is a non-reversible reaction, phenol formaldehyde cannot be softened again by heat.

This type of plastic is hard and rigid, but tends to be rather brittle in thin sections. However, it has a good electrical resistance, so it is not surprising that the 'bakelite' industry grew in conjunction with the electrical industries from the late nineteen-twenties onwards. Various mouldings for wireless-cabinets, motor-car parts, bottle-tops, switchgear, electric plugs, door-knobs, and a host of other articles are commonly made from bakelite. In most of these articles, a wood-flour filler is used; but, if greater strength is required, a fibrous filler (paper, rags, jute, or sisal) is employed.

23.71.2 *Urea-formaldehyde* plastics are basically similar to the phenol-formaldehyde types, since they depend upon the two-stage cross-linking reaction, but in this instance between molecules of urea and formaldehyde. The first stage of the reaction, however, results in the formation of a syrupy material. This is mixed with the filler, to give a moist, crumbly mass. After mixing with other reagents, this is allowed to dry out, producing the urea-formaldehyde moulding powder.

The fact that the urea-formaldehyde mixture passes through a syrupy stage makes it useful in other directions. For example, paper, cloth, or cardboard can be impregnated with it, and then moulded to the required shape, before being set by the application of heat and pressure. Weather-resistant plywood (24.71) can also be made, using urea-formaldehyde as the bonding material; whilst the syrup can also be used as a vehicle for colouring materials in the coating of metal furniture, motor cars, washing-machines, and refrigerators with a layer of enamel.

23.71.3 *Melamine-formaldehyde* plastics are also of similar type, but are much harder and more heat-resistant than either phenol-formaldehyde or urea-formaldehyde plastics. Moreover, they are more resistant to water, and consequently find use in the household, as cups and saucers, baths, and sundry kitchen utensils, particularly where greater heat-resistance is necessary.

23.72 *Polyester resins* of the thermoplastic type have already been mentioned (23.65). These are of the 'straight-chain' molecule form, which are incapable of linking up with other adjacent chain molecules; they are therefore thermoplastic. By using special monomers, however, polyesters can be produced with side-chains capable of forming cross-links between adjacent molecules, and these are therefore 'setting plastics', of which some are of the cold-setting type. For example, if styrene is added to the liquid polyester, it provides the necessary cross-links, and the liquid gradually sets without the application of any heat. Used with glass-fibre, to provide reinforcement, these materials are useful for building up such structures as the hulls of small boats, wheelbarrows, and car bodies.

Group	Compound	Relative density	Tensile strength (N/mm²)	Chemical resistance
Phenolics	Phenol formaldehyde ('bakelite')	1·45	50	Very good
	Urea formaldehyde	1·48	45	Fair
	Melamine formaldehyde	1·49	50	Good
Polyesters (setting types)		1·3	40	Fairly good
Alkyd resins		2·2	25	Fair
Polyurethanes		1·2	Mainly foams	Good
Epoxy resins	Heat-resistant type	1·15	70	Good
	General purpose	1·15	63	

Table 23.2—Properties and uses of thermosetting materials.

The glass fibre, impregnated with the liquid mixture, is built up a layer at a time on a suitable former. Pressure is not required, except to keep the material in position until it has set. These glass-fibre/polyester resin composites are very strong and durable.

23.72.1 *Alkyd resins* constitute a further group of polyesters. They are thermosetting materials originally introduced as constituents of paints, enamels, and lacquers. They have a high resistance to both heat and elec-

Safe working temperature (°C)	Relative cost	Typical uses
120	Low	Electrical equipment, radio-cabinets, vacuum-cleaners, ash-trays, buttons, cheap cameras, automobile ignition-systems, ornaments, handles, instrument-panels, advertising-displays, novelties and games, dies, gears, bearings (laminates), washing-machine agitators
80	Low	Adhesives, plugs and switches, buckles, buttons, bottle-tops, cups, saucers, plates, radio-cabinets, knobs, clock-cases, kitchen equipment, electric light-fittings, surface coatings, bond for foundry sand
130	Moderate	Electrical equipment, handles, knobs, cups, saucers, plates, refrigerator coatings, trays, washing-machine agitators, radio-cabinets, light-fixtures, lamp-pedestals, switches, buttons, building-panels, automotive ignition-blocks, manufacture of laminates
95	Moderate	Adhesives, surface coatings, corrugated and flat translucent lighting-panels, lampshades, radio grilles, refrigerator parts, Polyester laminates are used for hulls of boats, car bodies, wheelbarrows, helmets, swimming-pools, fishing-rods and archery-bows
230	Moderate	Enamels and lacquers for cars, refrigerators, and stoves. Electrical equipment for cars, light-switches, electric-motor insulation, television-set parts
120	Fairly high	Adhesives (glass to metal), paint base, wire-coating, gears, bearings, electronic equipment, handles, knobs. Foams are used for insulation, upholstery, sponges, etc. Rigid foams are used for reinforcement of some aircraft wings
200 / 80	Moderate	Adhesives (metal gluing), surface coatings, casting and 'potting' of specimens. Laminates are used for boat hulls (with fibre glass), table surfaces and laboratory furniture, drop-hammer dies. Epoxy putty is used in foundries, to repair defective castings

tricity, and are dissolved by neither acids nor many organic solvents. These resins are now available in powder form, which is generally compression-moulded (23.92.1). Liquid alkyds are used in enamels and lacquers for automobiles, stoves, refrigerators, and washing-machines.

23.73 *Polyurethanes* constitute a group of very adaptable plastics, comprising both thermoplastic and thermosetting materials. They are generally clear and colourless. One type is used in the manufacture of bristles, filaments, and films.

In another group of polyurethanes, carbon dioxide is evolved during the chemical process necessary for establishing cross-links between the chain molecules. This carbon dioxide is trapped by the solidifying polymer, thus producing a foam. The mechanical properties of the foam can be varied by using different materials to constitute the polymer. Thus, some of the foams become hard and rigid, whilst others are soft and flexible. Rigid thermosetting polyurethane foams are used as heat-insulators, and for strengthening hollow structures, since they can be poured into a space where foaming and setting will subsequently occur. Aircraft wings can be strengthened in this manner. Flexible sponges, both for toilet purposes and for seat upholstery, are generally of polyurethane origin.

23.74 *Epoxy resins* are produced in a similar way to polyesters; by being mixed with a cross-linking agent, which causes them to set as a rigid network of polymer molecules. They are used for manufacturing laminates, for casting, and for the 'potting' of electrical equipment. Excellent adhesives can be derived from the expoxy resins available as syrups, and they are particularly useful for metal glueing (26.20).

Solid exposy resins are sometimes mixed with phenolic resins for moulding purposes.

23.75 *Silicones.* The polymers mentioned so far in this chapter are based on the element carbon, and are known as organic compounds. Unfortunately, they all soften, decompose, or burn at quite low temperatures. However, more than a century ago, it was realised that there was a great similarity in chemical properties between carbon and the element silicon. Since then, chemists have been trying to produce long-chain molecules based on silicon, instead of carbon, hoping that in this way materials would be discovered which lacked many of the shortcomings of carbon polymers. Common compounds of silicon include quartz, glass, and sand —all substances which are relatively inert, and which remain unchanged after exposure to very high temperatures. Consequently, the more fanciful of our science-fiction writers have speculated on the possibility of a system of organic life, based on silicon instead of carbon, existing on the hot planets Venus and Mercury. Whilst their 'bug-eyed monsters' may remain a figment of the imagination, some progress has taken place here on Earth in producing polymers based on silicon, and called organo-silicon compounds, or 'silicones'.

Some of these compounds—of which sand is the basic raw material—have properties roughly midway between those of common silicon compounds, such as glass, and the orthodox organic plastics. Silicones are based on long-chain molecules in which carbon has been replaced by silicon and oxygen. They are available as viscous oils, greases, plastics, and rubbers. All are virtually non-combustible, and their properties remain constant over a very wide temperature-range. Thus, silicone lubricating oils retain their fluidity more or less unchanged at temperatures low enough to make ordinary oils congeal. Similarly, silicone rubbers remain flexible from low sub-zero temperatures up to temperatures high enough to decompose ordinary rubbers.

Silicones are water repellent, and are widely used for waterproofing clothes, shoes, and other articles; whilst silicone jelly is useful as a moisture-proof coating and sealing compound.

By forming still longer chain molecules, silicone plastics can be produced. These are very useful as an insulating varnish for electrical equipment designed to work at high temperatures. As moulding plastics, silicones are used in the manufacture of gaskets and seals for engineering purposes where high temperatures are involved, since they retain their plasticity and sealing efficiency under such conditions.

Silicone-resin paints provide durable finishes which clean easily, and which do not deteriorate appreciably. Such a finish applied to a motor car would normally outlast the car, and with a minimum of attention.

Mechanical properties of plastics

23.80 The strength of plastics is generally much lower than that of other structural materials. However, plastics are light substances, with a relative density between 0·9 and 2·0; so, when considered in terms of strength/weight ratio, they compare favourably with many metals and alloys. Figure 23.9 shows the types of stress-strain diagram obtained for different groups of plastics. It should be remembered, however, that strength varies considerably with the temperatures encountered during service. Thus, a thermoplastic material may have a tensile strength of, say, 70 N/mm^2 at 0°C, falling to 40 N/mm^2 at 25°C, and to no more than 10 N/mm^2 at 80°C.

23.81 The most important properties of plastics are their ability to withstand corrosion, and breakage by impact. Some materials, such as PTFE, have very low coefficients of friction, making them useful as dry bearings.

The main disadvantages of plastics are:

(1) their tendency to creep; that is, elongate slowly under load;
(2) their rapid deterioration at temperatures above 200°C; and
(3) the development of surface cracks in some polymers during service, this probably being due to the action of ultra-violet light.

Methods used to shape plastics

23.90 Most of the larger plastics manufacturers supply raw materials in the form of powders, granules, pellets, and syrups. These are purchased by the fabricators for the production of finished articles. As with metals, plastics can also be purchased in the form of extrusions and sheet, on which further work is to be carried out.

The principal methods of fabrication are moulding and extrusion, but raw materials obtainable in the liquid form may be cast.

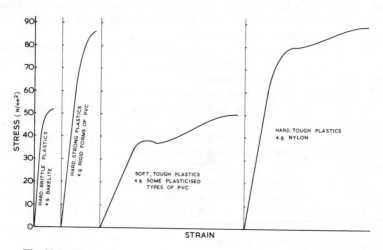

Fig. 23.9—Types of stress–strain diagram for different groups of plastics.

23.91 *Extrusion.* The principle of extrusion is shown in fig. 23.10. Here the plastic material is carried forward by the screw mechanism, and, as it enters the heated zone, it becomes soft enough to be forced through the die. The die aperture is shaped according to the cross-section required in the product.

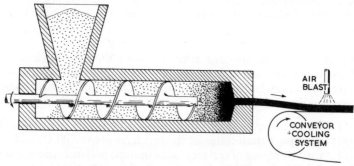

Fig. 23.10—The extrusion of plastics.
Often known as 'screw-pump' extrusion.

Man-made fibres are extruded using a multi-hole die, or 'spinneret'. Wire can be coated with plastics by extrusion, assuming that the machine is adapted so that the wire can pass through the die in the manner of a mandrel.

23.92 *Moulding.* Several important hot-moulding processes are commonly used.

23.92.1 *Compression-moulding* (fig. 23.11) is probably the most important, and is used for both thermoplastic and thermosetting materials, though it is particularly suitable for the latter. In either case the mould must be heated, but for thermoplastic substances it has to be cooled before the workpiece can be ejected. A carefully measured amount of powder is used, and provisions are made to force out the slight excess necessary to ensure filling of the mould cavity.

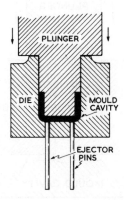

Fig. 23.11—One system of compression-moulding.

23.92.2 *Injection-moulding* (fig. 23.12) is a very rapid process, and is widely used for moulding such materials as polythene and polystyrene. The material is softened by heating it in the injection nozzle. The mould itself is cold, so that the plastic soon hardens and can be ejected. The workpiece generally consists of a 'spray' of components, connected by 'runners' which are subsequently broken off.

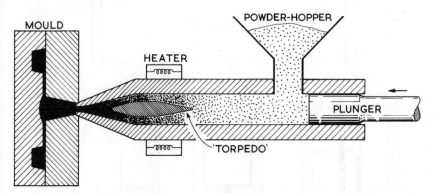

Fig. 23.12—The principle of injection-moulding.

23.92.3 *Transfer-moulding* is used for thermosetting plastics. The material is heated, to soften it, after which it is forced into the heated mould (fig. 23.13), where it remains until set. Rather more intricate shapes can be produced by this process than by compression-moulding.

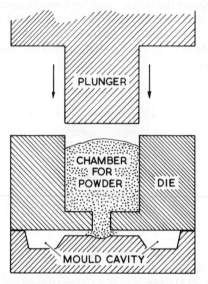

Fig. 23.13—Transfer-moulding.

23.92.4 *Blow-moulding* (fig. 23.14) is used to produce hollow articles. The plastic is first softened by heating, and is then blown by air pressure against the walls of the mould. In the diagram, extruded tube (parison) is being used.

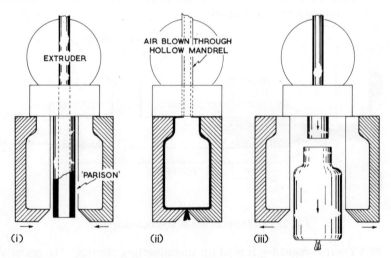

Fig. 23.14—Blow-moulding by interrupted extrusion.

23.92.5 *Vacuum-forming* is also used in producing simple shapes from thermoplastic materials in sheet form (fig. 23.15). The heated sheet is clamped at its edges, and is then stretched by the mould as it advances into position. The ultimate shape is produced by applying a vacuum, so that the work-piece is forced into shape by the external atmospheric pressure.

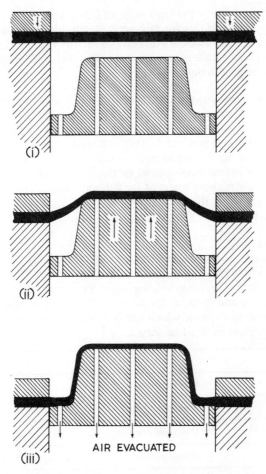

Fig. 23.15—Stages in a vacuum-assisted forming process.

23.93 *Casting* is limited to those plastics whose ingredients are obtained as liquids. The mixed liquids are poured into the open mould, and are allowed to remain at atmospheric pressure until setting has taken place.

Chapter Twenty-four
Wood and its Products

24.10 Roughly a third of the land mass of this planet is covered by trees. Unfortunately, most of this forest land is unsuitable for economic timber operations, or is not sufficiently accessible for the timber to be removed. Consequently, only a little over a quarter of the available forest land is used in actual timber production.

24.11 Wood is our oldest structural material, and is still employed in almost every industry. Much timber, generally of inferior quality, is also used in the production of paper pulp. This may be regarded as an excellent procedure when the end-product is this textbook, but one may feel less enthusiastic about the felling of trees to make possible the printing of the average Sunday newspaper—unless, of course, it ultimately finds partial redemption as a wrapping material for fish and chips. Trees are probably the most ancient living things on earth—some of them are more than 5000 years old—so perhaps we should treat them with due respect.

Types of wood
24.20 The forests of the world produce both softwoods and hardwoods. The former, which provide the bulk of timber for structural and joinery work, are coniferous, or cone-bearing trees. They grow mainly in northern Europe, Canada, and Asia, as far north as the Arctic Circle. Except for larch and southern cypress, which are deciduous trees, conifers are evergreens, with typical needle-like leaves.

HARDWOODS

Popular name	Botanical species	Origin
Ironbark	Eucalyptus paniculata	Australia
Ebony	Diospyros crassiflore	W. Africa
Greenheart	Ocotea rodiaei	S. America
Hickory	Carya sp.	N. America
Teak	Tectona grandis	S.E. Asia
Beech	Fagus sylvatica	Europe
Oak	Quercus sp.	Europe
Sapele	Entandrophragma cylindricum	W. Africa

Birch	*Betula* sp.	Europe
Iroko	*Chlorophore excelsa*	W. Africa
Walnut	*Juglans regia*	Europe
Plane	*Platanus acerifolia*	Europe
Wych elm	*Ulmus glabra*	Europe
Soft maple	*Acer rubrum*	N. America
Sycamore	*Acer pseudoplatanus*	Europe
English elm	*Ulmus procera*	Europe
African mahogany	*Khaya ivorensis*	W. Africa
American mahogany	*Swietenia macrophylla*	C. America
Grey poplar	*Populus cenescens*	Europe
White willow	*Salix viridis*	Europe
Obeche	*Triplochiton scleroxylon*	W. Africa

SOFTWOODS

Popular name	*Botanical species*	*Origin*
Pitch pine	*Pinus caribaea*	C. America
Douglas fir	*Pseudotsuga taxiflora*	W. America
European larch	*Larix decidua*	UK
Parana pine	*Araucaria augustifolia*	S. America
Scots pine	*Pinus sylvestris*	UK
Western hemlock	*Tsuga heterophylla*	N. America
European redwood	*Pinus sylvestris*	Europe
Canadian sitka spruce	*Picea sitchensis*	W. America
Quebec yellow pine	*Pinus strobus*	N. America
Canadian spruce	*Picea* sp.	W. America
European whitewood	*Picea abies*	N. America
Western red cedar	*Thuja plicata*	N. America

Table 24.1—Varieties of wood used commercially.

24.21 Many of the hardwoods are not particularly hard. Thus, the soft low-density balsa is classed as a hardwood, and the distinction between hardwoods and softwoods is botanical, rather than physical. Nevertheless, most of the common hardwoods are harder than the average softwoods. Hardwoods are generally deciduous, though a few are evergreens. All have broad leaves, and consequently a more complex wood structure to support this heavier load of foliage. Over one hundred different types of hardwood find commercial use, whereas little more than twenty species of softwood are commercially available. A selection of some of the better known types of wood is given in table 24.1. These woods are arranged roughly in order of strength and hardness.

The structure of timber

24.30 The trunk of a tree consists of two principal parts, the inner or 'heart-wood', which, like the skeleton of a mammal, gives it strength and rigidity, and the young active outer layers, called 'sapwood', which is a region of growth, and, like the flesh of a mammal, transmits and stores the food. The timber of living trees and freshly felled logs contains a considerable proportion of moisture; so, when the tree is felled, a period of 'seasoning' is necessary in order that loss of moisture, accompanied by shrinkage, can take place.

24.31 As a tree grows, it increases simultaneously in both height and girth. Growth is a seasonal process, and, in the case of trees grown in a temperate climate, a new layer forms each year. These layers are shown as a series of concentric 'growth-rings' in a cross-section cut through the trunk of a tree (fig. 24.1), and in this way its age can be estimated. A popular exhibit

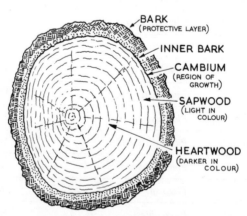

Fig. 24.1—Cross-section of a tree trunk.

in many natural history museums consists of a cross-section through the trunk of an ancient oak, important historic events of bygone days being marked on the appropriate growth-rings. Trees grown in some climates do not show these well defined growth-rings.

In a piece of metal, the building unit is the crystal; whilst a super-polymer is an agglomeration of large numbers of long, thread-like molecules. In living matter, both plant and animal, the simplest unit is the cell. As a tree grows, the wood tissue forms as long tube-like cells of varying shapes and sizes. These are generally termed 'fibres', and are arranged in roughly parallel directions along the length of the trunk (fig. 24.2).

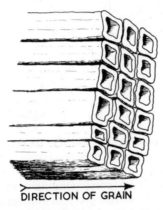

DIRECTION OF GRAIN

Fig. 24.2—The type of cellular structure in white pine ($\times$ 10 approx).

These cells vary in size from 0·025 to 0·5 mm in diameter, and from 0·5 to 5·0 mm in length, and are composed mainly of cellulose (23.40). The fibre-like cells are cemented together by the natural resin lignin; so wood can be regarded as a *naturally occurring laminate* (24.70). Wood contains approximately 60% cellulose, 28% lignin, and 12% sugars and other soluble materials. As the tree grows, the inner growth-rings die, and this portion acts only as a mechanical support or skeleton for the living outer layers.

24.32 Reference has already been made to 'fibre' in rolled metal sections (6.50), and it was stated that the maximum strength of such a material was in the direction of the fibres. The same principle applies in the case of wood, but to a much greater extent. Thus, European redwood may have a tensile strength of about 9 N/mm² in the direction of the grain; whilst 'across the grain' (i.e. perpendicular to it) the strength may be no more than 1·5 N/mm². It is therefore clear that the direction and slope of the grain or fibre plays a far more important part in timber than it does in metals. The slope of the grain within a piece of timber is very important, as indicated in fig. 24.3.

24.33 Timber has an excellent strength/weight ratio, and, when considered in this context, some types of wood can compete with steel. Unfortunately,

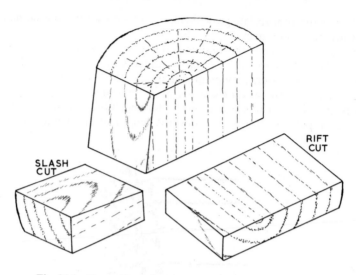

Fig. 24.3—Types of cut, in relation to the grain of the wood.

a bulky cross-section is necessary if wood is to sustain a tensile load; consequently, steel, with its correspondingly small cross-section, will generally be used instead. In compression and bending, provided there is no bulk disadvantage, wood may be preferable, as indeed it is for domestic building. Wood has a high modulus of elasticity, and, when vibrated, it produces strong resonance; hence it is used for the sounding-boards of many stringed musical instruments, such as violins and guitars—the traditional or 'acoustic' types, of course, not the electrically amplified devices manipulated by the vendors of pop music.

Moisture content and seasoning

24.40 Seasoning—or 'conditioning', as it is called in the trade—is the adjustment of the moisture-content of timber to a state of equilibrium with its surroundings. Timber shrinks when it dries, and expands when it absorbs moisture; so 'movement' one way or the other will occur when timber is put into surroundings containing more or less moisture than the equilibrium amount.

Moisture-content is expressed as a percentage of the *dry* weight; that is,

$$\text{moisture-content} = \frac{(\text{wet weight} - \text{dry weight})}{\text{dry weight}} \times 100\%$$

Some freshly felled timber of the more porous woods contains more than its own weight of moisture, i.e. the moisture-content is above 100%. After thorough air-drying, the moisture-content is reduced to about 18%, making the timber suitable for use when exposed to the outside air. If the timber is to be used in a centrally-heated building, it is necessary to reduce the moisture-content to 12%, or even less, and this can be achieved only by kiln-drying. Drying also prevents decay, as wood-rotting fungi will not attack timber with less than about 20% moisture-content.

As a result of shrinkage and loss of moisture, wood becomes stronger and more rigid, the strength in bending of most types increasing by about 50% from 'green' to dry.

24.41 Softwoods can be air-seasoned in about three to six months, depending upon the thickness and the weather conditions; whilst hardwoods such as oak or beech take about six months per centimetre of thickness. When kiln-drying is used, these times can be reduced to one week for softwoods, and four weeks for hardwoods of 50 mm thickness.

The preservation of timber
24.50 In 1628, the Swedish warship *Wasa* capsized and sank at the start of her maiden voyage. Until 1961, she remained at the bottom of Stockholm harbour, when it was discovered that she was in remarkably good condition, and was subsequently raised. This great oak ship, with all its beautifully carved decorations and wooden sculptures, has since had a museum built around it. The excellent state of preservation was explained by the fact that the water in which the wreck lay was too saline to support freshwater organisms, but not sufficiently salt to allow 'sea-worms' to survive. Thus, due to the absence of any predatory organisms, the wood did not decay.

Metals are subject to corrosion by fairly simple chemical attack in the presence of moisture, but timber, which is obtained from a living organism, is attacked by other living organisms—insects and fungi. Normally, when a tree dies, it is, in the course of time, reduced to its original chemical ingredients by the scavengers of the forest—boring insects, fungi, and bacteria. The chemical ingredients return to the soil, and so the cycle of nature repeats itself. When timber is used for constructional purposes, these scavengers are regarded as enemies, and steps must be taken to repel them.

24.51 *Insect pests.* In Britain, various types of beetle constitute the most important insects which attack wood. In each case, the female lays her eggs on or just beneath the surface of the wood, and the larvae (or grubs) which ultimately hatch eat their way through the timber until, considerably grown, they become dormant, and form pupae or chrysalids. Eventually, the adult beetles emerge, find their way to the surface via bore-holes, and the life-cycle repeats itself.

The *common furniture-beetle* attacks both softwoods and hardwoods, but seems to prefer old furniture. The *death-watch beetle*, on the other hand, generally attacks hardwoods, and is most frequently found in the roof timbers of old churches. It thrives in damp situations arising from a lack of ventilation, and derives its name from the tapping sound it makes during the mating season—a not unusual behaviour pattern among certain higher forms of life.

These two are the most common insect pests which attack wood, though there are several others encountered less frequently. They often reveal their presence by the little mounds of wood powder scattered around the scenes of their crimes, though the death-watch beetle differs from the others in that its larvae produce little bun-shaped pellets. If timber is badly infected, the only sure way to eradicate the pest is to cut out such timber and burn it. A number of proprietary preservatives are available to ward off such attack. To be effective, such preservatives should be permanent in their effect, poisonous to both insects and fungi, but non-toxic to human beings and animals.

24.52 *Fungus attack.* A very large number of fungi and moulds (or micro-fungi) attack timber in their search for food and lodgings. However, the best known are those which cause the effects generally known as *dry-rot* and *wet-rot*.

The fungus which causes dry-rot thrives in damp, poorly ventilated situations where a temperature between 16° and 20°C prevails. It consists of a somewhat disgusting, dark, feathery mass, with branching tendrils, and is able to penetrate brickwork. An unpleasant mouldy smell generally prevails when dry-rot is present. The timber becomes discoloured, and develops a dry, shrunken appearance. This is the origin of the term 'dry-rot', but it must be emphasised that the fungus which causes it operates only in damp situations.

The fungi responsible for wet-rot cause internal rotting of the wood, and thrive only in very wet conditions. A pale green scum first appears, but this soon changes to dark brown, and ultimately to black.

Whilst these forms of fungal attack occur most frequently in the basements and ground floors of buildings, they may occur elsewhere, if the combination of dampness and poor ventilation produce conditions under which they can thrive. All infected timber must be cut out and burnt, and the surviving sound wood treated with a preservative.

24.53 Wood-preservatives fall into three main groups:

(1) tar-oil derivatives (e.g. creosote)—applied by brushing, dipping, or by pressure;

(2) water-soluble materials—applied by pressure;

(3) materials in organic solvents—applied by brushing, dipping, or spraying.

Creosote is probably the most widely used wood-preservative. It is both economical and effective, and is suitable for the treatment of fencing, railway-sleepers, marine timbers, telegraph-poles, and wooden buildings. It does not attack metals, and is not washed out by rain. Its main disadvantage is that it is inflammable.

Engineering uses of timber
24.60 One speaks of English oak almost with reverence, recalling, as it does, the days of Drake, when 'Hearts of oak were our ships', and the building of an Empire had begun. Although sea-power now depends largely on a somewhat less glamorous steel hull, the uses of timber in general have become even more diverse in the many industries of the modern world.

24.61 *Building and civil-engineering industries.* These industries use roughly a third of the total wood products available. Timber has a fairly low strength and stiffness, when compared with metallic materials on the basis of equal sections; but, since it is also of low density, this permits members of relatively greater cross-section to be used. At the same time, since timber is a natural material, often grown under variable conditions, more deviation in its physical properties can be expected than is the case with a man-made product like steel, where properties can be controlled by standardised production methods.

The mechanical properties of timber are affected by its moisture-content, as well as by the number and spacing of defects such as knots, resin pockets, distorted grain, and pest damage. Timber is therefore graded according to the density of such defects, and grades have been introduced to to fit in with the British Standards Code of Practice CP 112: 1967, *The Structural Use of Timber*, so that stress-graded timber can be selected with a minimum of reject material.

The building industries use softwoods, the most important of which are European redwood, European whitewood, Western red cedar, Douglas fir, Western hemlock, and Canadian spruce.

24.62 *Mining industries.* Mining takes up about 12% of timber supplies in Great Britain. Although modern mechanised coal-face support systems have to some extent replaced the old-time pit-props, a large number of the latter are still used for supporting roofs. A variety of softwoods is employed for this purpose, including much small 'roundwood' from British forests. In addition, large quantities are imported from the USSR, Finland, and other Scandinavian countries.

24.63 *Transport industries.* About 10% of our timber supplies are used by the transport industries. British Rail has traditionally used large quantities of Scots pine, maritime pine, and Douglas fir as railway-sleepers, though this demand is diminishing as the use of concrete sleepers, in conjunction with welded-steel rails, develops.

Railway rolling-stock, road-transport vehicles, and shipping use large quantities of both softwoods and hardwoods, in many varieties.

24.64 *Other engineering uses.* These are many and varied. The foundry trades use considerable quantities of white pine, yellow pine, mahogany, and jelutong for pattern-making, whilst tool-handles account for considerable amounts of the hardwoods ash, beech, birch, box, hickory, hornbeam, and Tasmanian oak.

Laminated wood

24.70 Much progress in the use of timber as a constructional material has been made possible in recent years by the development of synthetic-resin adhesives (26.22). Glued lamination of timber ('gluelam') has many advantages over conventional methods of mechanical jointing, such as nailing, screwing, and bolting. Thin boards or planks can be bonded firmly together in parallel fashion, using these strong adhesives, and members be thus manufactured in sizes and shapes which it would be impossible to cut directly from trees. Timber members can be made indefinitely long by gluing boards end to end on a long sloping 'scarf' joint, and laminae can be 'stacked' to produce any necessary thickness and width. Moreover, thin laminae can be bent to provide curved structural members which are extremely useful as long-span arches, to cover churches, concert-halls, gymnasia, and similar large buildings. Arches a hundred metres or more long, with members up to two metres deep, are not uncommon. The main problem is in transporting them, since assembly 'on site' is unsatisfactory.

24.71 Adhesives used in glued laminations include casein and urea formaldehyde for interior work, and the more durable resins phenol formaldehyde and resorcinol formaldehyde for timbers exposed to weather, or other damp conditions. All synthetic-resin adhesives, many of which are available from 'do-it-yourself' stores, consist of two parts—the resin and an acid hardener, supplied separately or premixed in powder form. Setting time is often several hours, though special fast-setting hardeners are available.

Plywood, blockboard, and particleboard

24.80 An inherent weakness of natural timber is its directionality of properties, as shown by a lack of strength perpendicular to the grain. This fault is largely overcome in laminated board—or plywood, as it is commonly called—by gluing together thin 'veneers' of wood, so that the grain direction in each successive layer is perpendicular to that in the preceding layer. These veneers are produced by turning a steam-softened log in a lathe, so that the log rotates against a peeling knife. Most softwoods and many hardwoods can be peeled successfully, provided that the log is straight and cylindrical.

24.81 *Plywood* varies in thickness from 3 mm to 25 mm. The thinnest material is three-ply; whilst thicker boards are of five-ply or multi-ply

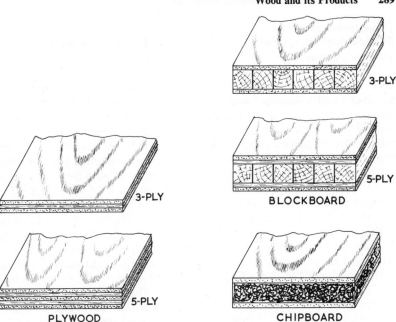

Fig. 24.4—Plywood, blockboard, and chipboard.

construction. An odd number of plies is always necessary, so that the grain on the face and the back will run in the same direction; otherwise the product will be unbalanced, and any change in the moisture-content during service may lead to warping.

24.82 *Blockboard* resembles plywood, in that it has veneers on both face and back. The core, however, consists of solid wood strips up to 25 mm wide, assembled and glued side by side (fig. 24.4). Adhesives used in the production of these materials are chosen to suit service conditions. Plywood and blockboard for interior use are glued with casein, which is resistant to attack by pests, but which will not withstand moisture for long periods. Consequently, plywood and blockboard destined for exterior use are bonded with urea-formaldehyde or melamine-formaldehyde resins. Better still, a mixture of phenol formaldehyde and resorcinol-formaldehyde is used to produce material which is 'weather- and boil-proof'.

24.83 Another important modern introduction in timber sheet material is *particleboard* or *chipboard*. This consists of a mass of wood chips, ranging from coarse sawdust to flat shavings, bonded with a resin, and sandwiched between veneers. Here the strength is dependent mainly on the resin bond, but there is no directionality of properties, since the particles constituting the wood filling are orientated at random.

Other timber products

24.90 Several other products of interest to us have their origin in wood as the raw material.

24.91 *Fibreboard.* Two widely used materials derived from timber are *hardboard* and *insulation board*, often referred to under the general title of *fibreboard*. Both are made from wood-pulp, in a similar manner. Small pulpwood logs are mechanically ground to a pulp, which is then fed, as a liquid mass, on to a mesh-belt conveyor, where it forms a layer some 50 mm thick. In the case of hardboard, some resin is incorporated into the pulp.

The mesh conveyor then carries the thick layer of pulp through a series of rolls, which compress it, and squeeze out most of the water. The compressed material is then cut into sheets, which are transferred to a multiple press equipped with heater platens. Hardboard is treated at higher pressures than insulation board, in order to give it a greater density. Finally, the pressed sheets are conditioned, to adjust the moisture-content. The most popular wood for the manufacture of these products is spruce, which produces the desired long fibres, but some fast-growing timbers like the South African *eucalyptus saligna* are also used.

Insulation board is similar in composition to hardboard, but it contains no resin, and is only lightly pressed, so that it retains the open texture which is essential to give it its lightness and insulation properties.

24.92 *Wood wool* is also derived from small roundwood. It is scraped from logs in long strands, which form a matted mass. This is then treated with cement, and compressed into slabs and blocks, to give a light material of high thermal insulation. These slabs and blocks are widely used for insulation in the internal walls and roofs of buildings.

Chapter Twenty-five
Asbestos, Ceramics, and Concrete

25.10 Materials based on asbestos and ceramics are of interest to the engineer in somewhat specialised fields. Both types of material are generally resistant to heat and to high temperatures, and also have a high electrical resistance, making many of them useful as insulators. Some ceramic materials are extremely hard and abrasion-resistant, whilst asbestos has an excellent resistance to attack by most common chemicals. Concrete is, of course, the most significant building material of the twentieth century. It associates rigidity, strength, and durability with low cost.

Asbestos and its derivatives
25.20 Asbestos is the name given to certain minerals which occur in nature in the form of long fibres, and which possess good heat-resisting properties. These minerals exist as a number of chemical compounds containing varying amounts of two or more of the following substances—magnesia, lime, iron oxide, alumina, and silica. The most important variety, which accounts for about 95% of the world's production, is a mineral called chrysotile, containing roughly equal parts of magnesia and silica.

25.21 All varieties of asbestos have high melting-points and low thermal and electrical conductivities. Some are more flexible than others, and can be woven into cloth or rope, whilst other varieties are more fire- and acid-resistant, making them more useful in filters, cements, and plastics. Thus, the commercial importance of asbestos is due both to its incombustibility and to its fibrous structure. It resists heat, flame, acids, and many other chemicals. It is like a mineral silk which remains undecayed by both time and weather.

25.22 It is thought that asbestos developed as the result of a combination of high pressure and temperature, following volcanic action millions of years ago. The main supplies of the mineral come from Canada, southern and central Africa, Russia, and the USA. After the mining operation, asbestos goes through a series of crushing and milling processes, in order to separate the fibres from the rock to which they are attached.

Properties of asbestos
25.30 Asbestos is a flexible, fibrous, silky material, from which fibres as fine as 1.8×10^{-6} mm in diameter can be separated. The fibres are strong, and, with suitable compositions, tensile strengths in the region of 1500 N/mm^2 can be obtained. For most types of asbestos, the strength begins

to fall at temperatures above 200°C, though chrysotile itself does not melt until a temperature of approximately 1500°C is reached.

Asbestos is the only naturally occurring mineral which can be woven into cloth, and resist flame, heat, time, weather, and many acids, alkalis, and other reagents.

Products consisting mainly of asbestos

25.40 These are products used mainly where flame-resistance, thermal and electrical insulation, or resistance to chemical attack are involved. They consist principally of asbestos to which small quantities of a suitable binder may be added to effect cohesion.

25.41 *Raw asbestos fibre* is used for insulation purposes, in particular as a packing for walls and floors, and in underground conduits. Since it resists corrosion, it is also used for the insulation of batteries.

25.42 *Asbestos textiles* are produced by spinning and weaving processes which are basically similar to those used for other fibres. Next to the asbestos-cement industry, probably the most significant use of asbestos is in the manufacture of yarn and cloth.

In addition to flame-, heat-, and acid-resistance, these textiles possess considerable durability. Consequently, they are used for such applications as belting for conveying hot material, padding and cloth covers in laundry-machines, as well as friction materials, industrial packing, and electrical and thermal insulation. Other textile products include braided and knitted materials; rope for caulking, seals, gaskets, and boiler expansion-joints; fibre for filtration; safety-curtains for theatres; fire-smothering blankets; fire-fighting suits; and many others.

25.43 *Asbestos paper* is manufactured by processes similar to those used for ordinary paper. A water-based pulp is first produced; this is spread evenly on a travelling wire mesh, and successively 'de-watered' by pressure application. Finally, the consolidated mass is dried and smoothed.

The product has many uses, both industrial and domestic. A large quantity is employed as electrical insulation, mainly in the form of wrapping-tape. As thermal insulation and for fire protection, it is used in the lining of filing-cabinets, military helmets, automobile silencers, and armoured-car roofs. On the domestic scale, it is used as a thermal insulation in cookers and radiator covers, and for baking-sheets and table-mats.

25.44 *Asbestos millboard* is virtually a thick form of asbestos paper. It is a rigid form of insulation, which cuts and drills easily, and can be nailed or screwed, whilst at the same time withstanding temperatures up to 550°C without serious deterioration.

It has many uses, such as the linings of stoves, ovens, and other heaters; fire-proof linings of electric switch-boxes, safes, and doors; heat insulation in incubators; fire-proof wall-board; and, on a smaller scale, table-mats and pads.

Composite materials containing asbestos

25.50 Many engineering materials have a composite structure; that is, they contain two or more different substances, which are blended together homogeneously, but which remain as separate particles or fibres in the structure. Products as diverse in their properties as cemented-carbide tools (13.41) and concrete (25.70) fall into this category.

In most cases, a composite material is produced to give extra strength. Thus, plastics may be reinforced with cloth, glass-fibre, or carbon fibres; whilst an increase in rigidity is achieved by using ceramic substances in materials called 'cermets' (25.62). Finally, a cheap filler-material (or 'aggregate') can be used to increase the bulk of a product at low cost, as in concrete.

Asbestos fibre is used as a reinforcement for Portland cement, the fibre behaving in much the same way as the steel rods in reinforced concrete. As a filling material for plastics, it will increase strength, toughness, electrical insulation, and chemical- and heat-resistance.

25.51 *Asbestos-cement products* contain from 10% to 75% Portland cement, according to their application. The strong asbestos fibres which they also contain strengthen the products in the same way as do the steel rods in reinforced concrete. The saving in load, however, as compared with reinforced concrete, can be as much as 70 or 80% where roofing and other structural members are involved.

The cement and asbestos fibre are usually mixed dry, and water is then added, the slurry subsequently being compressed between rollers to render it dense. Alternatively, the slurry is fed into a suitably shaped mould, and pressure is applied to express excess water, and increase the density of the product, which is then allowed to set and harden for 24 to 48 hours.

Typical uses of flat asbestos-cement sheet include roofs and walls of small buildings, partitions, portable buildings, fire-protection booths, laboratory table-tops, fireproof layers on insulated boards, linings of gas or electric cookers, switch-boards, and many others. Asbestos-cement pipes are used for carrying water, sewage, and gas; and are now widely used for such purposes in ordinary domestic installations.

25.52 *Asbestos felt* and papers, impregnated with asphalt, serve similar purposes to ordinary roofing-felts which are made from vegetable fibres. The asbestos-based material has a much longer life, and is suitable for flashings and damp courses in walls. Roofing-sheets for houses and other buildings may be manufactured from asbestos felt, in cases where thermal insulation is important, as well as resistance to weather.

25.53 *Magnesia-asbestos insulation* is used both for fire-proofing and for thermal insulation. It contains about 85% magnesium carbonate, along with asbestos fibre, which acts as a binder and reinforcement material. To this mixture is added sufficient water to allow treatment in a filter-press, where the material is pressed into the required shape. The product is then dried out in a steam-heated kiln, for about three days.

The resulting material has a very low density, because of the large proportion of air-filled cavities it contains. Consequently, its thermal conductivity is very low, making it very useful for the manufacture of the outer insulation layers of furnaces, and insulation-jackets for drying-ovens, pipes, and locomotives.

25.54 *Asbestos reinforced and filled plastics.* These include both thermoplastic and thermosetting materials (23.21), to which asbestos has been added in the form of textiles, papers, felt, yarns, or loose fibre. The 'plastics' used vary in composition, from resin and shellac to phenolic substances. The former are used when a material of high electrical insulation and reasonable strength is required, whilst the latter will result in a product combining high strength and shock-resistance with excellent chemical- and heat-resistance. Many laminates are manufactured in which asbestos mats or textile sheets are sandwiched in a suitable plastic substance.

Typical applications of asbestos/plastic materials include electrical-circuit breakers, castors, pulleys, pipes, roller- and sleeve-bearings, and washing-machine agitators. On the more spectacular scale, asbestos-reinforced phenolic plastics have been used in missile technology, particularly in nose-cones, fins, thermal-insulation barriers, and exhaust systems. Auxiliary fuel-tanks produced by the British Aircraft Corporation are made in asbestos/phenolic resin materials.

25.55 *Friction materials* containing asbestos are used extensively in the automobile industry, in the form of clutch- and brake-linings. These materials have properties which cannot easily be obtained with other materials. They do not overheat excessively, since the natural slipperiness of the asbestos gives rise to suitable braking action; consequently, they are not subject to excessive wear. If a small amount of cotton fibre is incorporated in the mixture, this chars to carbon during use, and produces a polished working-surface.

The substances used to impregnate and bond these linings include asphalts, various oils (both vegetable and mineral), phenolic and furan resins, casein, coal tar, and rubber. Rubberised linings are used extensively, and moulded compounds contain 30–50% rubber with 50–70% asbestos fibre, though the latter may be mixed with varying amounts of other filler materials.

Ceramics
25.60 The term 'ceramic' is derived from the Greek *keramos*, meaning 'potter's clay'. Gradually this meaning was extended to include all pro-

ducts made from fired clay, such as bricks, tiles, fireclay refractories, and electrical porcelain, as well as pottery tableware. Many substances now classed as ceramics in fact contain no clay, though all are relatively hard, brittle materials of mineral origin, with high fusion-temperatures. Thus materials like glass, vitreous enamel, and hydraulic cement are now included under the general heading of ceramics; whilst a number of metallic oxides such as alumina, beryllia, zirconia, and magnesia form the basis of high-temperature ceramics.

Clay is really a thermosetting material (23.21). In the raw state, its molecules are arranged in layers which are separated by molecules of water. This allows the layers to glide over each other, leading to the characteristic plasticity of unfired clay. Firing drives off the water molecules, and, at the same time, strong chemical links form between the layers, so that the clay becomes hard and rigid. Since the change is not reversible, clay cannot be softened again once fired, and it is thus a thermosetting material.

As far as engineering purposes are concerned, the main features which make ceramics useful are:

(1) refractoriness, or the ability to withstand high temperatures without deterioration,
(2) strength and rigidity at high temperatures,
(3) freedom from creep (4.50) at high temperatures, and
(4) hardness, and resistance to wear.

25.61 *High-temperature ceramics.* Gas-turbine and turbo-jet blades made from ceramic materials can be used at higher temperatures than those

Substance	Chemical name	Melting-point (°C)	Characteristics and uses
Alumina	Aluminium oxide	2050	Widely used in sparking-plugs, cutting-tools, crucibles, pyrometer-sheaths, gauges (with a life twenty times that of steel), surface-plates for precision-checking equipment
Beryllia	Beryllium oxide	2350	Crucibles for special materials, as a moderator in high-temperature nuclear reactors
Magnesia	Magnesium oxide	2800	A very refractory material, furnace-linings and crucibles
Thoria	Thorium oxide	3050	Not used much, because it is 'fissionable'
Zirconia	Zirconium oxide	2690	Used in the 'stabilised' form, liners for jet and rocket motor tubes, facing high-temperature furnace walls

Table 25.1—Some high-temperature ceramics.

made from the most refractory metallic alloys. Materials which can be used include alumina, beryllia, magnesia, zirconia, thoria, and spinel; as well as multi-component substances such as silicon carbide bonded with clay, beryllia bonded with clay, high-alumina porcelains, and high-magnesia porcelains.

Similar materials are used as a coating to protect surfaces from deteriorating as a result of contact with high temperatures. The ceramic shrouding of parts in rocket motors consists of tiles composed of zirconia, beryllia, silicon carbide, or zirconium boride. Such compounds will withstand temperatures in excess of 2000°C.

25.61.1 *Tool materials.* Whilst many of the better-known 'hard metal' tool materials are of the 'cermet' type, described below, some cutting tools are derived from a more-or-less pure ceramic base. These substances possess great hardness and good compressive strengths, even at high temperatures; though, compared with metals, their tensile strength is low, and tools made from them are relatively brittle. Ceramic tools can be used at higher cutting speeds than can 'hard metals', but they are particularly useful in cutting tough materials such as plastics, rubber, and wood, as well as metals. The more important 'ceramic bodies' used for this purpose include sintered high-purity alumina (corundum), boron carbide, and recrystallised silicon carbide (carborundum); whilst alumina bonded with glass and other materials is also used. Grinding wheels are usually composed of abrasive particles of alumina or silicon carbide, bonded with such materials as clay, quartz, or feldspar.

25.62 *Cermets.* Although ceramics generally have high melting-points, and are reasonably strong at high temperatures, most of them are rather brittle materials, with poor resistance to mechanical and thermal shock. Metals, on the other hand, are temperature- and shock-resistant, but generally lack strength and rigidity at high temperatures. Hence, a useful combination of properties is often achieved by mixing metal and ceramic in suitable proportions.

Powder-metallurgy methods (7.40) are often used for this purpose. Powders of suitable particle size are thoroughly mixed in the correct proportions, and are then compressed in dies, at high pressures. A degree of cold-welding occurs between the metal particles in the mixture, and, if the resultant 'compact' is then sintered (heated to a temperature above that of recrystallisation for the metal), grain-growth occurs across the boundaries of the minute welds, and knits together the metal particles, giving a tough and continuous structure. Hard rigid ceramic particles are then bonded in a tough metallic matrix.

If a material of considerable rigidity at high temperatures is required, then the proportion of ceramic may need to be high, in order that the resultant cermet will not deform (fig. 25.1). On the other hand, if tensile strength combined with stiffness and rigidity is required, then reinforce-

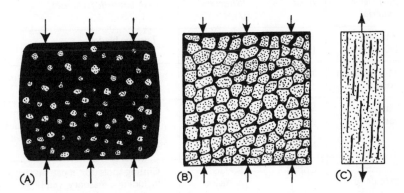

Fig. 25.1—Composite materials.
(A) Low ceramic-content—the cermet deforms in compression.
(B) High ceramic-content—the cermet withstands compressive loads and is also tough.
(C) Here the metal fibres improve the tensile strength, but the material is also rigid, due to the ceramic body.

ment of the ceramic with fibres or rods of the metal may be desirable, as is the case with reinforced concrete. When the metal-content is high—say 65–75%—the overall properties are nearer to those of a metal, giving good ductility and toughness, but poor strength and a high rate of creep (4.50) at high temperatures. With a high ceramic-content, the cermet tends to be relatively brittle, but has a higher tensile strength and a lower creep-rate at high temperatures.

Cermets generally are suitable for such uses as lamp filaments; aircraft jet-engine parts; gas-turbine parts; rocket-engine components; cutting-, drilling-, and grinding-tools; friction parts; nuclear-power applications; heating-elements; bearings; and magnetic-core materials (table 25.2).

Particle material	Bond	Structural type	Characteristics and uses
Flake silver or copper	Graphite	Laminated	Current brushes—low friction
Alumina (70%)	Chromium (30%)	Bonded particles	Very suitable for high-temperature service. Good resistance to impact and thermal shock
Magnesia	Nickel	Flame-sprayed heat-resistant coating	Applicable to stainless steels, alloy steels and Inconel—to raise the working temperature by about 80°C

Table 25.2—Some cermets.

Particle material	Bond	Structural type	Characteristics and uses
Alumina (40–70%)	Iron (30–60%)	Bonded particles	Turbine blades
Molybdenum boride	Nickel	Bonded particles	Cutting-tools for machining titanium
Titanium carbide	Various alloys containing Mo, Al, and Cr	Bonded particles	In aircraft engines where refractoriness, thermal shock-resistance and resistance to oxidation are necessary
Tungsten carbide or Titanium carbide	Cobalt	Bonded particles	Cutting-tools for most materials, including masonry and glass; metal-forming dies

Table 25.2—Some cermets—*cont'd.*

25.62.1 *Tool materials* of the cermet type are well known. They consist of extremely hard, abrasion-resistant particles, held together by a strong, shock-resistant bonding material. The most widely used and best known consists of particles of tungsten carbide bonded with tough, strong cobalt (fig. 25.2), though the carbides of titanium or tantalum are sometimes used instead of tungsten carbide.

Fig. 25.2—The type of structure in a tungsten carbide/cobalt cermet. Particles of tungsten carbide (white) in a cobalt matrix (black) (× 1000).

25.62.2 *Filament-reinforced ceramics* are cermets of the type shown in fig. 25.1c, in which a hard, rigid ceramic body is reinforced with strands of a tough, shock-resistant metal. By this definition, reinforced concrete falls into this class, but, at a more sophisticated level, ceramic materials used to protect rocket nozzles are reinforced with a network of tungsten wire. Similarly, metal grids of wire or tape are used to anchor a ceramic layer to a base plate. Zirconia reinforced with tantalum or molybdenum alloys is used in ram-jets, plasma-chambers, and heat-exchangers.

Cement and concrete

25.70 A number of different types of cement exist, but possibly the best known, and certainly the most widely used, is Portland cement. In the dry state, this is a greenish-grey powder, produced by firing a mixture of limestone (or chalk) and clay at a temperature high enough to cause a chemical reaction resulting in the formation of calcium silicates and aluminates. When mixed with water, Portland cement 'sets'. The chemistry of the setting process is complex, and probably not fully understood; however, the reaction between water and the silicates and aluminates produces a hard, rigid mass.

Concrete is produced from a plastic mixture of stone aggregates and cement paste. When hard, it has many of the characteristics of natural stone, and is extensively used in building and civil engineering. In the 'wet' state, it can be moulded easily, and is one of the cheapest constructional materials in Britain, because of the availability of suitable raw material, and also the low maintenance costs of the finished product.

The aggregate may be selected from a variety of materials. Stone and gravel are most widely used, but in some cases other substances available cheaply, such as broken brick and furnace slag, can be employed. Sand is also included in the aggregate, and, in order to obtain a dense product, a correct stone/sand ratio is essential. Aggregate materials should be clean, and free from clay.

The proportion of cement to aggregate used depends upon the strength required in the product, and varies from 1:3 in a 'rich' mixture to 1:10 in a 'lean' one. The cement/aggregate ratio commonly specified is 1:6, and this produces excellent concrete, provided that the materials are sound, and properly mixed and consolidated. Other things being equal, the

Fig. 25.3—A satisfactory cement structure.

The aggregate should consist of different grades (or sizes) in the correct proportions. In this way, the smaller particles fill in the spaces between the larger ones, whilst particles of sand occupy the remaining gaps. The whole is held together by a film of cement.

greater the size of the large stones, the less cement is required to produce concrete of a given strength.

The cement and its aggregates are mixed in the dry state, water is then added, and the mixing process is continued until each particle of the aggregate is coated with a film of cement paste. The duration of the hardening period is influenced by the type of cement used, as well as by other factors such as the temperature and humidity of the surroundings. Concrete made with Portland cement will generally harden in about a week, but rapid-curing methods will reduce this time. Concrete structures can be 'cast *in situ*' (as in the laying of foundations for buildings), or 'pre-cast' (sections cast in moulds, and allowed to harden before being raised into their final positions).

25.71 *Plain concrete.* This is suitable for use in retaining walls, dams, and other structures which rely for their stability on great mass. Other uses include foundations where large excavations have to be filled; whilst vast quantities of plain concrete are employed in modern motorway construction. At the other end of the scale, many small pre-cast parts are manufactured; for example, 'reconstituted stone' for ornamental walls, as well as sundry other articles of garden 'furniture'.

25.72 *Reinforced concrete.* In common with many ceramic materials, concrete is stronger in compression than in tension. For structural members such as beams, where stresses are both compressive and tensile, the use of plain concrete would be most uneconomical, as indicated in fig. 25.4 (i) and (ii). In reinforced concrete, the tensile forces are carried by fine steel rods. Full advantage can then be taken of the high compressive strength of the concrete, and the cross-section can be reduced accordingly. The rods are so shaped (fig. 25.4 (iii)) that they are firmly gripped by the rigid concrete. Thus, reinforcement allows concrete members to be used in situations where plain concrete of adequate strength would be unsuitable because of the bulk of material necessary.

25.73 *Pre-stressed concrete.* One possible source of weakness in reinforced concrete is illustrated in fig. 25.4 (iii), which indicates the formation of cracks in the side of the beam which is in tension. Steel has a much greater elasticity (4.21 and 6.10) than concrete; thus, though the stretched steel rods are still well within their safe limits of stress, the concrete, being less elastic, will have developed a large number of cracks. One way of counteracting this is to utilise the elastic properties of the steel to give an initial *compression* to the concrete; that is, to 'pre-stress' it.

A steel cable is stretched elastically, and, whilst in tension, a concrete beam is cast around it (fig. 25.5 (i)). When the concrete has hardened completely, the *external* tension at the free ends of the cable is removed (fig. 25.5 (ii)). Immediately, the cable attempts to shrink to its original length (since it has been stressed within its elastic limit), and, in so doing,

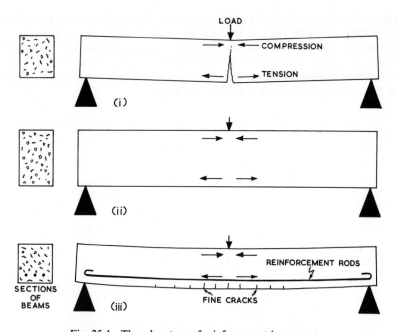

Fig. 25.4—The advantage of reinforcement in concrete.

In (i), a plain concrete beam fails at the edge which is in tension. A beam strong enough in this respect (ii) will be uneconomically bulky. In (iii), steel reinforcement-rods support the portion of the beam which is in tension.

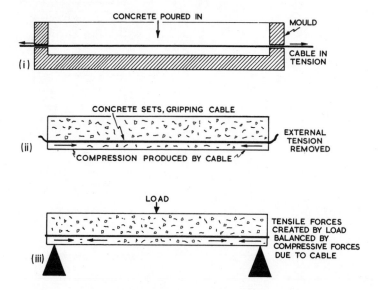

Fig. 25.5—The principles of pre-stressed concrete.

exerts a force of compression on the concrete. Provided that the cable is in the correct position relative to the cross-section of the beam, these compressive forces produced in the concrete will exactly balance the tensile forces caused by the applied load when the beam is in service (fig. 25.5 (iii)). In the fully loaded condition, therefore, there should be no resultant tension in the concrete, and cracks will not develop.

Chapter Twenty-six
Methods of Joining Materials

26.10 Metal welding, as carried out by the blacksmith, is without doubt one of the most ancient of metal-working processes; yet the modern technology of welding has undergone revolutionary change during the last two or three decades. In many instances where expensive riveting processes were once employed, steel sheets are now successfully joined by welding. The fabrication of the American 'Liberty' ships, during the Second World War helped to establish this technique.

26.11 Many methods are used for joining materials. Wood is screwed or nailed, as well as being joined by glue. For many purposes, metals are still riveted, or joined by bolts and nuts. However, in this chapter we shall consider only those methods of joining materials where a 'continuous' joint is employed; that is, adhesion is produced by forces of attraction acting between the fundamental particles of the materials involved.

Glues and adhesives
26.20 Adhesion signifies the joining together of two dissimilar materials, and is dependent upon forces of attraction which operate between molecules in the surface of the adhesive and molecules in the surfaces of the materials being joined. Generally speaking, the larger the molecules involved, the greater are these forces of attraction, and hence the greater the adhesion. Consequently, most adhesives are organic compounds, being composed of very large and complex molecules.

26.21 Another important feature is that a successful adhesive must mould itself perfectly to the surfaces being joined, in order that its molecules will remain in close contact with the molecules of the surfaces being joined. Adhesives bond two 'substrates'* together either by chemical attraction for the surfaces involved, or by a mechanical intertwining action, by which the adhesive is carried *into* the two substrates, and sets as a tough film. Adhesives used in industry generally work on a combination of both methods, and they normally consist of solutions or suspensions. In these forms, the adhesive material can conveniently be applied to a substrate. To facilitate quick setting of the adhesive, the problem of removing the water (or other solvent carrying the adhesive) as rapidly as possible remains. The simplest method is to use direct heat for the purpose. Unfortunately, this means that, to heat the glue-line, the whole assembly must be heated,

---

* This term signifies the surface and those regions near to the surface of the work-piece.

Group	Adhesive substance (or raw material)	Materials joined
Animal glues	Animal hides or bones Fish Casein (from milk) Blood albumen	Wood, paper, leather, and fabrics
Vegetable glues	Starch Dextrines	Paper and fabrics
	Soya beans	Paper-sizing
Natural resins	Bitumens (including asphalts)	Felt, laying floor-blocks
	Gum arabic	Paper and fabrics
	Shellac	Glass, metals
Inorganic cements	Sodium silicate	Foundry moulds
	Ceramic compositions, such as Portland cement and plaster of Paris	
Elastomer materials	Natural rubber and its derivatives	Rubber, sealing-strips (automobile industries), gaskets
	Synthetic rubbers	Nylon, footwear industries
Thermo-plastic synthetic resins	Vinyls, e.g. polyvinyl acetate	Wood furniture assemblies
	Cellulose derivatives, e.g. cellulose nitrate	Glass, paper, balsa-wood models
	Acrylics, e.g. polymethyl methacrylate	Optical elements
	Polyamides	Paper and leather
	Polystyrene	Polystyrene
Thermo-setting synthetic resins	Phenol, urea, melamine, and resorcinol formaldehydes	Weather-proof plywoods
	Epoxies	Optics, glass, polystyrene, nylon, metals, etc. Wide range of uses
	Polyurethanes	Paper, fabrics, etc.
	Furanes	Foundry moulding, phenol, urea, and melamine formaldehydes
	Silicones	Polythene, PTFE, silicone rubbers

Table 26.1—Glues and adhesives.

and then cooled before further handling; a decrease in production-rate would result.

Radio-frequency waves are a means of applying heat only to the point where it is required, namely the glue-line. By means of very high-frequency vibrations acting on the molecules of the adhesive, and trying to vibrate them, considerable heat is generated by the friction caused by molecules resisting the process. This energy is converted to heat, and dries off the solvent from the adhesive.

The heating process has a further very important effect with some adhesives of the polymer type. 'Cross-linking' (23.45) between adjacent molecules takes place, as, for example, it does in phenolic resins during moulding. This reaction strengthens the actual adhesive layer, and often increases its resistance to heat and water.

Table 26.1 lists some of the more important substances used as adhesives. Often these materials are blended to produce a better adhesive, particularly in the case of synthetic resins and rubbers. For example, well known adhesives like 'Bostik' and 'Evo-stick' are basically rubber-resin mixtures.

26.22 Service requirements must be considered in the choice of an adhesive. The most important requirements include strength, temperature-range, and resistance to water or moisture. The working properties of the adhesive are also important, and include the method of preparation and use, the storage-life, drying-time, odour, toxicity, and cost of bonding. Many adhesives require some type of solvent, in order to make them fluid: the solvent should evaporate reasonably quickly. Other adhesives, such as animal glues, need to be heated to make them fluid.

As with other joining processes, surface preparation is important. Surfaces must, above all, be thoroughly degreased; whilst the removal of dust and loose coatings is essential. Sand-blasting, wire-brushing, or grinding may have to be used.

Soldering and brazing
26.30 A solder must 'wet'—that is, alloy with—the metals to be joined, and, at the same time, have a freezing-range which is much lower, so that the work itself is in no danger of being melted. The solder must also provide a mechanically strong joint.

26.31 Alloys based on tin and lead fulfil most of these requirements. Tin will alloy with iron and with copper and its alloys, as well as with lead; and the joints produced are mechanically tough. Suitable tin–lead alloys melt at temperatures between 183°C and 250°C, which is well below the temperatures at which there is likely to be any deterioration in the materials being joined. Best-quality tinman's solder contains 62% tin and 38% lead, and, being of eutectic composition (9.11), melts at the single temperature of 183°C. For this reason, this alloy will melt and solidify quickly at the lowest possible temperature, passing directly from a completely liquid to a completely solid state, so that there is less opportunity for a joint to be broken by disturbance during soldering.

Because tin is a very expensive metal as compared with lead, the tin content is often reduced to 50%, or even less. Then the solder will freeze over a range of temperature, between 183°C and approximately 220°C. Solders are sometimes strengthened by adding small amounts of antimony.

26.32 For a solder to 'wet' the surfaces being joined, the latter must be completely clean, and free of oxide films. Some type of flux is therefore used to dissolve such oxide films, and expose the metal beneath to the action of the solder. Possibly the best-known flux is hydrochloric acid ('spirits of salts'), or the acid zinc chloride solution which is obtained by dissolving metallic zinc in hydrochloric acid. Unfortunately, such mixtures are corrosive to many metals, and, if it is not feasible to wash off the flux residue after soldering, a resin-type flux should be used instead. Aluminium and its alloys are particularly difficult to solder, because of the very tenacious film of oxide which always coats the surface.

26.33 Brazing is fundamentally similar to soldering, in that the jointing material melts, whereas the work-pieces remain in the solid state during the joining process. Brazing is used where a stronger, tougher joint is required, particularly in alloys of higher melting-point. Most ferrous materials can be brazed successfully. A borax-type flux is used, though, for the lower temperatures involved in silver soldering, a fluoride-type flux may be employed.

Composition %							Freezing-range (°C)	Type and uses
Sn	Pb	Sb	Ag	Cu	Zn	Cd		
62	38	—	—	—	—	—	183	Tinman's solder
50	50	—	—	—	—	—	183–220	Coarse tinman's solder
43·5	55	1·5	—	—	—	—	188–220	General-purpose solder
12	80	8	—	—	—	—	243–250	For soldering iron and steel
—	—	—	—	50	50	—	870–880	For brazing iron and steel
—	—	—	80	16	4	—	740–795	High-grade silver solders for use on brass, copper, 'Monel', and stainless steels
—	—	—	50	15	17	18	625–635	
—	—	—	10	52	38	—	820–870	Low-grade solder

Table 26.2—Solders, silver solders, and brazing solders.

Arc-welding processes

26.40 Early arc-welding processes made use of the heat generated by an arc struck between carbon electrodes, or between a carbon electrode and the work. 'Filler' metal to form the actual joint was supplied from a separate rod. The carbon arc is now no longer used in ordinary welding processes, and has been replaced by one or other of the metallic arc processes.

26.41 *Metallic-arc welding*, using hand-operated equipment, is the most widely used fusion-welding process. In this process, a metal electrode serves both to carry the arc and to act as a filler-rod which deposits molten metal into the joint. In order to reduce oxidation of the metal in and around the weld, flux-coated electrodes are generally used. This flux coating

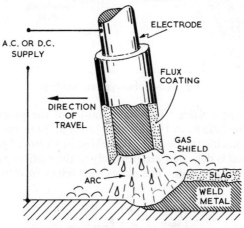

Fig. 26.1—Metallic-arc welding.

consists of a mixture of cellulose materials, silica, lime, calcium fluoride, and deoxidants such as ferro-silicon. The cellulose material burns, to give a protective shield of carbon dioxide around the weld; whilst the other solids combine to form a protective layer of fusible slag over the weld. Either a.c. or d.c. may be used for metallic-arc welding, the choice depending largely upon the metals being welded.

26.42 *Submerged-arc welding* is essentially an automatic form of metallic-arc welding which can be used in the straight-line joining of metals. A tube, which feeds powdered flux into the prepared joint, just in advance of the electrode, is built into the electrode holder. The flux covers the melting end of the electrode, and also the arc. Most of the flux melts, and forms a protective coating of slag on top of the weld metal. The slag is easily detached when the metal has cooled.

The process is used extensively for welding low-carbon, medium-carbon, and low-alloy steels, particularly in the fabrication of pressure-vessels, boilers, and pipes, as well as in shipbuilding and structural engineering.

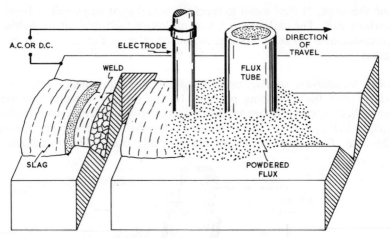

Fig. 26.2—Submerged-arc welding.

26.43 *Electro-slag welding.* The main feature of the process, which was developed in the USSR, in 1953, is that heavy sections can be joined in a single run, by placing the plates to be welded in a *vertical* position, so that the molten metal is delivered progressively to the vertical gap, rather as in an ingot-casting operation. The plates themselves form two sides of the

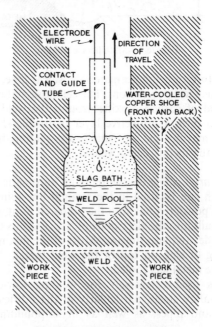

Fig. 26.3—Electro-slag welding.

'mould'; whilst travelling water-cooled copper shoes dam the flow of weld metal from the edges of the weld, until solidification is complete (fig. 26.3).

In this process, the arc merely initiates the melting process, and thereafter heat is generated by the electrical resistance of the slag, which is sufficiently conductive to permit the current to pass through it from the electrode to the metal pool beneath.

The process was originally developed for joining large castings and forgings, but its use has been extended to cover many branches of the heavy engineering industry. It can be used for welding a wide range of steels, and also titanium.

26.44 *Gas-shielded arc-welding*. Gases such as nitrogen and carbon dioxide, which are often referred to as being 'inert', will, in fact, react with some molten metals. Thus, nitrogen will combine with molten magnesium; whilst carbon dioxide will react with steel, oxidising it under some circumstances. The only truly inert gases are those which are found in small quantities in the atmosphere, viz. argon, helium, neon, krypton, and xenon. Of these, argon is by far the most plentiful—comprising 0.9%, by volume, of the atmosphere—and is therefore used for filling electric light bulbs, and also as a protective atmosphere in inert-gas welding. In the USA, substantial amounts of helium are derived from natural gas deposits; so it is used there as a gas-shield in welding. Since argon and helium are expensive to produce, carbon dioxide (CO_2) is now used in those cases where it does not react appreciably with the metals being welded.

In gas-shielded arc-welding, the arc can be struck between the workpiece and a tungsten electrode, in which case a separate filler-rod is needed, to supply the weld metal. This is referred to as the 'tungsten inert gas' (or TIG) process. Alternatively, the filler-rod itself can serve as the electrode, as it does in other metallic–arc processes. In this case, it is called the 'metallic inert gas' (or MIG) process.

26.44.1 *The TIG process* was developed in the USA during the Second World War, for welding magnesium alloys, other processes being unsatisfactory because of the extreme reactivity of molten magnesium. It is one of the most versatile methods of welding, and uses currents from as little as 0.5 A, for welding thin foil, to 750 A, for welding thick copper, and other materials of high thermal conductivity.

Since its inception, it has been developed for welding aluminium and other materials, and, because of the high quality of welds produced, TIG welding has become popular for precision work in aircraft, atomic engineering, and instrument industries.

26.44.2 *The MIG process* uses a consumable electrode which is generally in the form of a coiled, uncoated wire, fed to the argon-shielded arc by a motor drive. Most MIG welding equipment is for semi-automatic

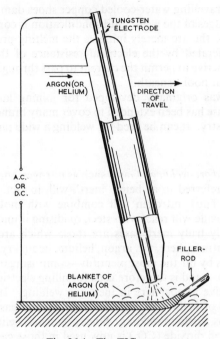

Fig. 26.4—The TIG process.

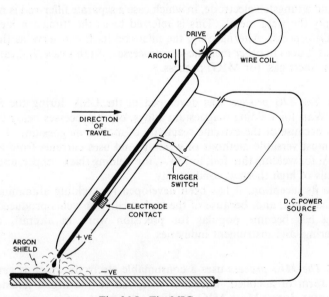

Fig. 26.5—The MIG process.

operation, in which the operator guides the torch or 'gun' (fig. 26.5), but has little else to do once the initial control settings have been made.

In addition to the advantages arising from semi-automatic operation, the MIG process is generally a clean welding process, due to the absence of flux. Consequently, the MIG process is one of the most diversely used welding methods, in terms of the number of different jobs with which it can cope successfully. Some industries where it finds application include motor-car manufacture, shipbuilding, aircraft engineering, heavy electrical engineering, and the manufacture of tanks, pressure-vessels, and pipes.

26.44.3 *The CO_2 welding process* is reasonably effective, provided the filler-rod used contains adequate deoxidising agents to cope with any oxidation which may arise due to the weld metals reacting with carbon dioxide. The normal CO_2 process is a modified MIG process, in which argon has been replaced by carbon dioxide, and in which the electrode wire is rich in deoxidants.

26.45 *Plasma-arc welding.* In this relatively new process, a suitable gas, such as argon, is passed through a constricted electrical arc. Under these conditions, the gas ionises; that is, its atoms split up into electrons and positively charged particles the mixture being termed 'plasma'. As the ions recombine, heat is released, and an extremely hot 'electric flame' is produced. The sun's surface consists essentially of high-temperature plasma, though even higher temperatures of up to 15 000°C can be produced artificially.

The use of plasma as a high-temperature heat source is finding application in cutting, drilling, and spraying of very refractory materials, like tungsten, molybdenum, and ceramics, as well as in welding.

Other fusion-welding processes
26.50 Some welding processes make use of heat obtained by a chemical reaction. Oxy-acetylene welding is probably the best known of these, though the Thermit process also continues to have its uses. In addition, some sophisticated methods based on both electron and laser beams have been developed in recent years.

26.51 *Oxy-acetylene welding.* Formerly, gas-welding ranked equally important with metallic-arc methods, but the introduction of argon-shielding to the latter process placed gas-welding at a disadvantage for welding metals where a flux coating is necessary. Consequently, the use of oxy-acetylene welding has declined in recent years. However, it is still widely used where maintenance or general repair work is involved.

Both oxygen and acetylene can conveniently be stored in cylinders, and their flow to the torch can easily be controlled by using simple valves. Although acetylene normally burns with a smoky, luminous flame, when oxygen is fed into the flame in the correct proportions, an intensely hot flame is produced, allowing temperatures in the region of 3500°C to be attained. Such a high temperature will quickly melt all ordinary metals, and is necessary in order to overcome the tendency of sheet metals to conduct heat away from the joint so quickly that fusion never occurs.

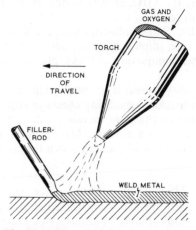

Fig. 26.6—The principles of gas-welding.

As with most other fusion-welding processes, a flux-coated filler-rod is used to supply the weld metal.

26.51.1 *Bronze-welding* is the term applied to the joining of metals of high melting-point, such as mild steel, by the use of copper-alloy filler-metals. An oxy-acetylene flame is generally used to supply the heat. Bronze-welding differs from true welding in that little or no fusion of the work-pieces takes place.

26.52 *Thermit welding* is used chiefly in the repair of large iron and steel castings. A mould is constructed around the parts to be joined, and above this is a crucible containing a sufficient quantity of the Thermit powder. The parts to be welded are first preheated, and the powder then ignited.

Thermit powder consists of a mixture of powdered iron oxide and aluminium dust, in calculated proportions. The chemical affinity of aluminium for oxygen is greater than the affinity of iron for oxygen; hence the powders react so:

iron oxide + aluminium = aluminium oxide + iron + heat

The heat of reaction is so intense that molten iron is produced, and this is tapped from the crucible into the prepared joint.

26.53 *Electron-beam welding.* When a stream of fast-moving electrons strikes a target, the kinetic energy of the electrons is converted into heat. Since the electron-beam can be focused sharply, to impinge at a point, intense heat is produced there. To be effective, the process must be carried out in a vacuum, otherwise electrons tend to collide with molecules of oxygen and nitrogen present in the air-space between the electron-gun and the target.

Consequently, electron-beam welding is at present used mainly for 'difficult' metals which melt at high temperatures, such as tungsten, molybdenum, and tantalum; and also for the chemically reactive metals beryllium, zirconium, and uranium, which benefit from being welded in a vacuum.

26.54 *Laser welding.* The term 'laser' is an acronym; that is, a name coined from the initial letter of each important word of its descriptive title. In this case, 'laser' represents Light Amplification by Stimulated Emission of Radiation. The device consists of a suitable generator of light pulses—or 'optical pump'—and an active element, and it emits a beam of light of extremely high intensity.

At present, laser technology is probably still in its infancy, but already it has had great successes in the fields of medical research, and in space technology. In the field of metal manufacture, it will ultimately find application in cutting, drilling, shaping, and welding, particularly in *micro* spot-welding.

Pressure-welding processes

26.60 In order to weld together two pieces of iron, the blacksmith first heats them to a high temperature, and then brings them into contact under pressure. He does this by hammering them together on an anvil, thus perpetuating one of the most ancient of metallurgical processes used. Whilst it is highly probable that Tubal Cain used this process some 6000 years ago (Genesis, iv, 22), we know with certainty that smith-welding was carried out by the Ancient Greeks, at least 400 years BC.

Apart from smith-welding, which is still used widely in many different forms, modern pressure-welding processes are mostly of the electrical-resistance type, in which the passage of a heavy electric current generates sufficient heat to permit welding.

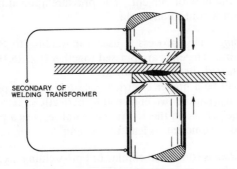

SECONDARY OF
WELDING TRANSFORMER

Fig. 26.7—Spot-welding.

26.61 *Spot-welding*. In this process, the parts to be joined are overlapped, and firmly gripped between heavy metal electrodes (fig. 26.7). An electric current of sufficient magnitude is then passed, so that local heating of the work-pieces to a plastic state occurs. Since the metal at that spot is under pressure, a weld is produced. This method is used principally for joining plates and sheets, but in particular for providing a temporary joint.

26.62 *Projection-welding* is a modified form of spot-welding, in which the current flow, and hence the resultant heating, are localised to a restricted area, by embossing one of the parts to be joined. When heavy sections have to be joined, projection-welding can be used where spot-welding would be unsuitable because of the heavy currents and pressures required. Moreover, with projection-welding, it is easier to localise the heating to that zone near the embossed projection. As fig. 26.8 shows, the projection in the upper work-piece is held in contact with the lower work-piece (i). When the current flows, heating is localised around the projection (ii), which ultimately collapses under the pressure of the electrodes, to form a weld (iii).

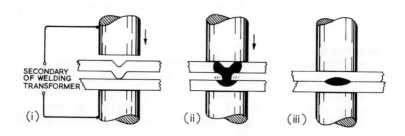

Fig. 26.8—Projection-welding

26.63 *Seam-welding* resembles spot-welding in principle, but produces a continuous weld by using wheels as electrodes (fig. 26.9). The work-pieces are passed between the rotating electrodes, and are heated to a plastic state by the flow of current. The pressure applied by the wheels is sufficient to form a continuous weld.

26.64 *Butt-welding*. This process is used for welding together lengths of rod, tubes, or wire. The ends are pressed together (fig. 26.10), and an electric current is passed through the work.

Since there will be a higher electrical resistance at the point of contact (it is most unlikely that the two ends will be perfectly square to each other), most heat will be generated there. As the metal reaches a plastic state, the pressure applied is sufficient to lead to welding.

26.65 *Flash-welding* is somewhat similar to butt-welding, except that heat is generated by striking an arc between the two ends which are to be joined.

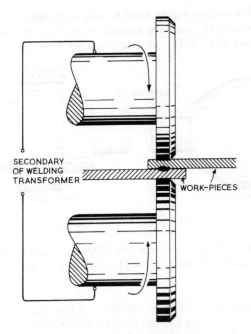

Fig. 26.9—Seam-welding.

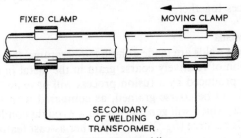

Fig. 26.10—Butt-welding.

Not only are the ends heated in this way, but any irregularities there are melted away. The ends are then brought together quickly, under pressure, so that a sound weld is produced.

26.66 *Other pressure-welding processes.* Several other welding processes which involve the use of pressure are worth mention here.

26.66.1 *Induction-welding.* In this process, the parts to be joined are pressed together, and an induction coil is placed around the joint. The high-frequency current induced in the work heats it to welding temperature.

26.66.2 *Friction-welding* can be used to join work-pieces which are in rod form. The ends of the rods are gripped in chucks, one rotating, and the other stationary (fig. 26.11). As the ends are brought together, heat is

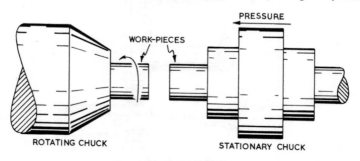

Fig. 26.11—Friction-welding.

generated, due to friction between the slipping surfaces. Sufficient heat is generated to cause the ends to weld. Rotation ceases as welding commences.

26.66.3 *Cold-pressure welding*, in the form of lap-, seam-, and butt-welds, generally involves making the two surfaces slip relative to each other under great pressure. Any oxide films are broken, and a cold-weld forms between the two surfaces. Spot-welds are produced in this way by punching the two work-pieces together.

Structures of welds
26.70 Since most welds are made at high temperatures, this inevitably leads to the formation of relatively coarse grain in the metal in and around the weld. A weld produced by a fusion process will have an 'as-cast' type of structure, and will be coarse-grained, as compared with the materials of the work-pieces, which will generally be in a wrought condition. Not only will the crystals be large (fig. 26.12), but other as-cast features, such as the segregation of impurities, may be present, giving rise to intercrystalline weakness.

When possible, it is of advantage to hammer a weld *whilst it is still hot*. Not only does this smooth up the surface, but, since recrystallisation of the weld metal will follow this mechanical working, a tough fine-grained structure will be produced. Moreover, the effects of segregation will be significantly reduced.

Some alloys are difficult to weld successfully, because of structural changes which accompany either the welding process or the cooling which follows. Thus, air-hardening steels, particularly high-chromium steels, tend to become martensitic and brittle in a region near to the weld; whilst 18–8 austenitic stainless steels may ultimately exhibit the defect known as 'weld-decay' (13.52), unless they have been 'proofed' against it.

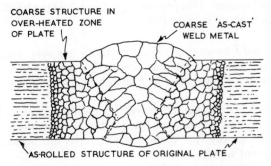

Fig. 26.12—The crystal structure of a fusion-weld.

Welding of plastics

26.80 Thermoplastic materials can be welded together by methods which are fundamentally similar to those used for welding metals. All methods depend upon the application of heat, and sometimes pressure.

26.81 *Hot-gas welding.* In this process, a jet of hot air from a welding torch is used (fig. 26.13). In other respects, this method resembles oxy-acetylene welding of metals. A thin filler-rod is used, and is of the same material as

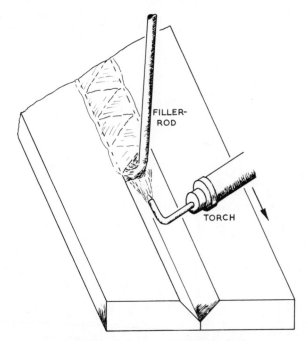

Fig. 26.13—The hot-gas welding of plastics.

that of the work-pieces. This process is used widely in the building of chemical plant, particularly in rigid PVC.

26.82 *Seam- and spot-welding.* These methods are very similar to those used in metal-joining, except that, since plastics are non-conductors of electricity, the 'electrodes' must be heated by high-frequency coils.

26.83 *Stitch-welding* is a method of welding thermoplastics, using a device similar to a sewing-machine, but fitted with two electrodes, which weld the material progressively. Again, the electrodes are heated by a high-frequency current supply.

26.84 *Jig-welding.* In this method, the work-pieces are gripped in jigs. These jigs act as electrodes which are heated by high-frequency fields.

26.85 *Friction-welding* also known as spin-welding, is similar in principle to the process which has recently been developed for joining metals (26.66.2). One part is rotated at speed, whilst the other is pressed against it. This heats both of the work-pieces. Motion is then stopped, and the parts are held together until the joint hardens.

Chapter Twenty-seven
The Choice of Materials and Processes

27.10 It is rare for the designer engineer to be given a completely free hand in selecting his material and the type of process by which he will shape it. Only when national prestige is at stake—as in the production of inter-planetary rocketry—or when a state is waging a war in defence of its existence, is the engineer asked to give of his best, regardless of cost.

Everyday engineering production, on the other hand, involves highly competitive enterprise, which, in the long run, nearly always leads to some sacrifice of quality, in favour of lower cost. The manufacturer is not neces-sarily to blame for this state of affairs, since he is usually satisfying an existing demand, even though that demand is often stimulated by a vigorous advertising-campaign. Often the cost of perfection is too high, and in most cases the effects of competition, coupled with a lack of power of discrimination by the consumer, lead to the manufacture of an article which is 'cheap and nasty'—or in some cases just nasty.

27.11 It would be possible to design and produce a *family* motor car having a reasonably trouble-free life of twenty years or so. Naturally, this would cost rather more in terms of materials—rust-resisting bodywork, for example —than is spent on normal production models. Some degree of 'craftsman-ship' would also need to replace mass-production methods. But even though such an automobile were produced at, say, double the cost of the familiar 'tin lizzie', one suspects that there would be few 'takers'. The type of consumer who could afford such a car generally prefers to change his car every two years or so. Whether the reason for this is 'Jonesmanship', 'because he likes to change', or because he is disenchanted with his present model matters little in this context, but the net result is that expensive retooling and reorganisation, particularly in respect of body presswork, must be paid for—by the consumer—at frequent intervals.

27.12 Public demand, whether contrived by the manufacturer or not, affects the philosophy of engineering production in many other lines in a similar manner. Thus, washing-machines, domestic cookers, vacuum-cleaners, and the like all come under the influence of 'Jonesmanship' and 'planned obsolescence'. Vast numbers of such machines are scrapped long before they have reached the end of their useful life. This seems to be a needlessly wasteful aspect of the affluent society, but it is within the terms imposed by this economic climate that we must consider the choice of engineering materials.

27.13 In choosing a material for a specific application the engineer must consider:

(1) the ability of the material to withstand service conditions,
(2) the method(s) by which it will be shaped, and
(3) the overall cost of both material and process(es).

In some cases, a number of materials may be satisfactory in respect of fulfilling the service requirements. Normally the engineer will choose the one which, when the cost of forming and shaping is taken into account, results in the lowest overall production cost. Thus, the cheapest material may not necessarily be the one which is used; a more expensive material may be capable of being formed very cheaply, and so give a lower overall production cost.

27.14 The service requirements of a material may involve properties which fall under one or all of the following three headings:

Mechanical properties	*Physical properties*	*Chemical properties*
(a) strength	(a) relative density	(a) resistance to oxidation
(b) elasticity	(b) melting-point or softening-temperature	(b) resistance to other forms of corrosion
(c) toughness	(c) thermal conductivity	
(d) stiffness	(d) coefficient of expansion	
(e) hardness	(e) effect of temperature changes on properties	
(f) fatigue resistance	(f) electrical conductivity	
(g) resistance to creep	(g) magnetic properties	
(h) frictional properties		

27.15 As an example, let us suppose we have to produce very small diameter wire, with the highest possible electrical conductivity. Since small diameter wire is required, it follows that material of very high ductility should be used, provided, of course, that it also has the necessary high electrical conductivity. It just so happens that copper and silver are extremely ductile metals in the pure state, and that their electrical conductivities are also highest when the metals are pure. Although silver has the highest electrical conductivity of all metals, copper, with well over 90% of the electrical conductivity of silver, is a very much less expensive metal. Consequently, pure copper has always been the traditional metal for electrical purposes.

27.16 During recent years, however, the position has changed, due to a big increase in the price of copper causing users to look elsewhere for a cheaper substitute. Aluminium has roughly 50% of the specific conductivity of copper, but is also much less dense. Thus, taken *weight for weight* (fig. 17.2), aluminium is a better conductor of electricity than is copper. This may be important in some circumstances, and, indeed, for many years aluminium has been used instead of copper for high-tension cables in the grid system. However, since high-frequency currents tend to be conducted by the 'skin' of a conductor, aluminium wire coated with copper has recently come into use where it is necessary to keep the diameter of the wire *small*, as in transformer-windings and the like. In this case, the extra production costs involved in coating the aluminium wire with copper are justified when the alternative is to use pure copper, which is now very expensive.

Many other cases come to mind. For example, the desirability of using a rust-resisting material for the manufacture of automobile bodywork has been mentioned. The cost of stainless steel, however, is very high as compared with that of mild steel. Moreover, since stainless steel work-hardens quickly, high forming costs would be incurred, again as compared with mild steel. For these reasons, mild steel coated with paint is used, although, at the time of writing, it seems as though greater use of plastics such as ABS can be expected in the not too distant future.

27.17 Tensile strength is a property which is of paramount importance in many fields of engineering design; yet quite often it is the ratio of strength/weight (or more properly strength/relative density) which has to be used in assessing the suitability of a material. This is particularly true in the aircraft industry. Though there is often very little to choose between strength/weight ratios for many of the common engineering alloys, it is often the alloy of low relative density which proves to be the more useful, since the greater bulk of material allows a more *rigid* cross-section to be designed. Consequently, aluminium alloys are more useful than steels in aircraft design, because the greater cross-section of the aluminium alloy, as compared with that of steel (for the same load-carrying capacity), enables a more complex shaped cross-section to be produced, and this will be much more rigid under the action of forces tending to bend the member.

The metal beryllium would be extremely useful as a material in aircraft construction, since it has a very high strength/weight ratio. Unfortunately, it is an extremely scarce and expensive metal, in addition to being difficult to fabricate, and these factors preclude its use in any but the more specialised applications.

27.20 The properties of a material which affect its suitability for forming and shaping operations include:

(*a*) malleability
(*b*) ductility
(*c*) castability
(*d*) machinability
(*e*) strength
(*f*) capacity, if any, for heat-treatment
(*g*) methods of joining available.

Thus, forging processes can be carried out on materials which are malleable in either the hot or cold state. Ductile materials can be drawn or deep-drawn, spun or stretch-formed. Some materials are neither malleable nor ductile, and can be produced only in the cast state. If great accuracy of dimensions is required in such a component, then an expensive investment-casting process may be involved.

27.21 Very often the choice of shaping process will depend upon the number of components required. Thus it may be desirable to use a die-casting process to produce a component, and thus eliminate machining operations which would otherwise be necessary if sand-casting were employed instead. Unfortunately, the use of a die-casting process will only be economical if a large number of castings, say 5000 or more, is required, thus covering the high initial cost of the steel dies. Similarly, drop-forging is only economical if very large numbers of components are required, since expensive die-blocks must be sunk. Sometimes a compromise may be struck, at the expense of some reduction in mechanical properties, by using a malleabilised-iron sand-casting in cases where only a few components are required.

Frequently, the shaping process is also used to develop strength in a component. Thus, an aluminium cooking-pot or the bodywork of an automobile become sufficiently rigid as a result of work-hardening which accompanies the shaping process.

27.22 Some metals and alloys have a very limited scope for shaping. The material for gas-turbine blades, which needs to be strong and stiff at high temperatures, is, for this very reason, difficult to forge, even at very high temperatures. Shaping is thus an expensive matter, and investment-casting is often the answer to the problem in such cases.

27.23 Conditions are continually changing, as new materials become available. During the last two decades, plastics have replaced metals and glass for innumerable applications. As an example, most curtain-rail material was extruded from a 60–40 type brass. Now much of it is produced in plastic, which is extruded on to a steel core. Many components which were once made in the form of metal die-castings are now manufactured as plastic mouldings, die-casting still being retained as the shaping process where higher strength, rigidity, and temperature-resistance are required. The corrosion-resistance of a plastic moulding is often very much higher than was that of the former die-casting, particularly if the latter was made from an alloy which was in a poor state of purity.

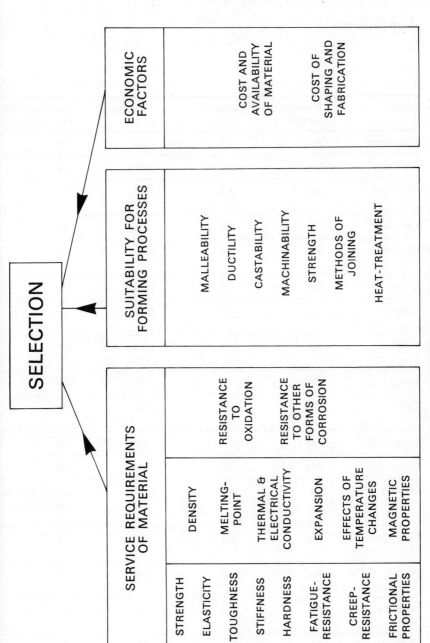

Fig. 27.1—The Selection of Materials.

Appendices

Metal	Chemical symbol	Relative density	Melting-point (°C)	Tensile strength (N/mm²) (soft)
Aluminium	Al	2·7	660	59
Antimony	Sb	6·6	630	10
Beryllium	Be	1·8	1285	310
Cadmium	Cd	8·6	321	80
Chromium	Cr	7·1	1890	220
Cobalt	Co	8·9	1495	250
Copper	Cu	8·9	1083	220
Gold	Au	19·3	1063	120
Iron	Fe	7·9	1535	500
Lead	Pb	11·3	327	18
Magnesium	Mg	1·7	651	180
Manganese	Mn	7·2	1260	500
Mercury	Hg	13·6	−39	molten at ordina: temperatures
Molybdenum	Mo	10·2	2620	420
Nickel	Ni	8·9	1458	310
Niobium	Nb	8·6	1950	270
Platinum	Pt	21·4	1773	130

Appendix I—Properties of the important engineering metals.

ngation *)*	*Characteristics and uses*
	The most widely used of the non-ferrous metals
	A brittle, crystalline metal, used in limited amounts in bearing and type-metals
3	A light metal, the use of which is limited by its scarcity. Used in beryllium bronze
	Used for plating steel, and to strengthen copper for telephone-wires
	A metal which resists corrosion, and is therefore used for plating, and in stainless steels
	Used mainly in permanent magnets and in super-high-speed steels
	Now used mainly where very high electrical conductivity is required; also in brass and bronzes
	Of little use as an engineering metal, because of softness and scarcity. Used mainly in jewellery, and as a system of exchange
	Quite soft when pure, but rarely used in engineering in the unalloyed form
	Not really the densest of metals, as the phrase 'as heavy as lead' suggests. Very resistant to corrosion—used in chemical engineering
	Used in conjunction with aluminium in the lightest of engineering alloys
	Very similar to iron in many ways—used mainly as a deoxidant in steel
olten at dinary mperatures	The only liquid metal at normal temperatures
	A heavy metal, used mainly in alloy steels. One of the main constituents of modern high-speed steels
	A very adaptable metal, used in both ferrous and non-ferrous alloys. The metallurgist's main 'grain-refiner'
	Also known as 'columbium' in the USA. Used mainly in alloy steels, and as an 'atomic energy' metal
	A precious white metal. Used in scientific apparatus, because of its high corrosion-resistance; also in some jewellery

Metal	Chemical symbol	Relative density	Melting-point (°C)	Tensile strength (N/mm²) (soft)
Silver	Ag	10·5	960	140
Tin	Sn	7·3	232	11
Titanium	Ti	4·5	1725	230
Tungsten	W	19·3	3410	420
Uranium	U	18·7	1150	390
Vandium	V	5·7	1710	200
Zinc	Zn	7·1	420	110
Zirconium	Zr	6·4	1800	220

Appendix I—Properties of the important engineering metals—*cont.*

ongation	*Characteristics and uses*
	Has the highest electrical conductivity, but is used mainly in jewellery and, in a few countries, for coinage
	Widely used but increasingly expensive. 'Tin cans' carry only a very thin coating on mild steel
	A fairly light metal, which is becoming increasingly important as its price falls due to the development of its technology
	Used in electric lamp filaments, because of its high melting-point. Is also the main constituent of most high-speed steels
	Now used mainly in the production of atomic energy
	Used in some alloy steels
	Used widely for galvanising mild- and low-carbon steels. Also as a basis for some die-casting alloys. Brasses are copper-zinc alloys
	Used as a grain-refiner in steels. It is also used for atomic energy applications

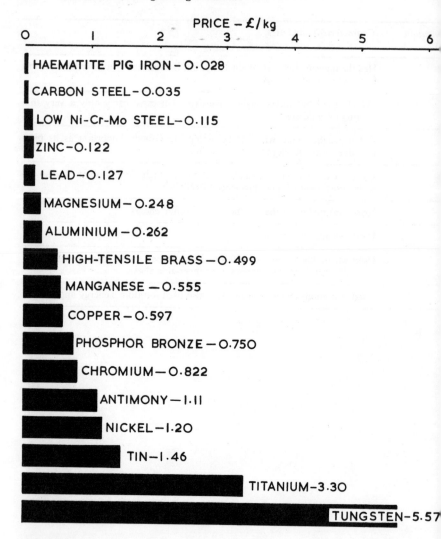

Appendix II—The relative prices of metallic engineering materials in terms of cost per unit mass (1972).

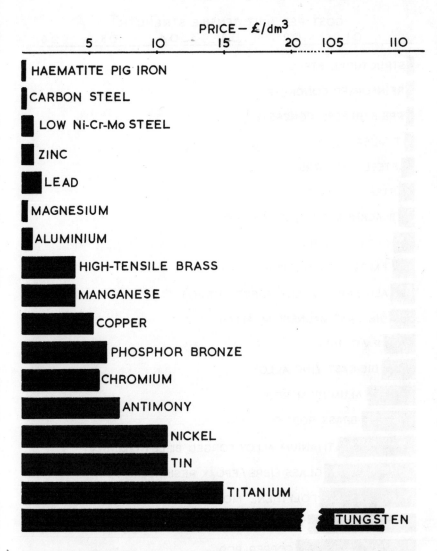

Appendix III—The relative prices of metallic engineering materials in terms of cost per unit volume (1972).

COST PER UNIT TENSILE STRENGTH †

0·1	0·2	0·3	0·4	0·5	0·6

STRUCTURAL STEEL

REINFORCED CONCRETE

PRE STRESSED CONCRETE

TIMBER

STEEL FORGINGS

STEEL CASTING

BLACKHEART MALLEABLE IRON

GREY CAST IRON

LM.22 CAST ALUMINIUM ALLOY

ALUMINIUM ALLOY SHEET (HS.30)

DIE-CAST MAGNESIUM ALLOY

P V C PIPE

DIE-CAST ZINC ALLOY

ALUMINIUM ROD

BRASS ROD

TITANIUM ALLOY FORGED BILLET (IMI.679)

GLASS FIBRE / EPOXY RESIN

POLYTHENE MOULDINGS

TITANIUM ALLOY ROD (IMI.318)

COPPER ROD

TITANIUM ALLOY SHEET (IMI.318)

† THESE FIGURES ARE PRICES (IN £ PER METRE LENGTH) OF BARS MADE OF SUFFICIENT DIAMETER SO THAT THEY WILL JUST SUPPORT A MASS OF I 000 kg.

Appendix IV—The relative prices of some common engineering material in terms of cost per unit tensile strength (1972).

Index